ONSITE WASTEWATER TREATMENT SYSTEMS

Bennette Day Burks
Mary Margaret Minnis

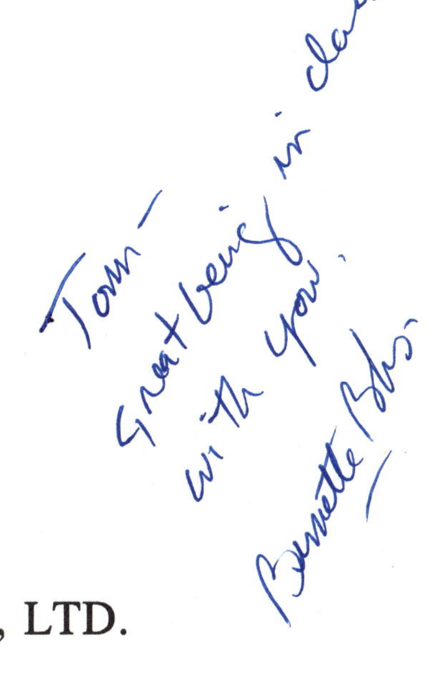

HOGARTH HOUSE, LTD.
Madison, WI

Printed in the United States of America

Publisher's Cataloging-in-Publication Data
Burks, Bennette D.
Onsite wastewater treatment systems / Bennette Day Burks, Mary Margaret Minnis
p. cm
Included bibliographical references and index.
1. Sewage—Purification I. Minnis, Mary Margaret.
II Title
TD745.B8 1994

Library of Congress Catalog Card Number: 94-76249
ISBN 0-9641049-0-3: $49.95 Softcover

Published by:
Hogarth House, Limited
15 Burchard Lane
Rowayton, CT 06853

Fourth printing, revised, January 2002
10 9 8 7 6 5 4

Table of Contents

FOREWORD

Our first involvement with onsite wastewater treatment systems began like many people's: we each purchased houses served by traditional septic systems. Unlike most, however, we each had scientific backgrounds and training in wastewater management. What surprised us was the dearth of information on onsite wastewater treatment systems (or "onsite systems") despite the fact that onsite systems serve about 25 percent of the population of the United States. More surprising and contrary to much of modern engineering practice, the administration of onsite system regulations is bound by tradition, myths, and secondary considerations that make advancement of the technology difficult. Literature available for the typical homeowner is limited to hints and tips. Until recently, engineering texts devoted one page or two on the topic (which students were expected to read on their own initiative). The academic literature is scattered. Perhaps of greater concern, regulations governing the installation and use of onsite systems vary from community to community and provide a host of contradictory requirements that frustrate the legitimate manufacturer, conscientious regulator, and well-intentioned user.

Given this frustration, this text is intended to provide bridges: bridges between onsite wastewater treatment systems and municipal wastewater treatment systems, bridges between manufacturers and consumers, between consumers, and regulators, and between regulators and manufacturers; and bridges between the academics and practitioners. The text also explains why onsite system regulations can be difficult to change.

We had four audiences in mind as we wrote the text: science or engineering students, regulators, policy makers, and interested citizens. We have provided theory where we think theory on onsite systems is lacking but otherwise tried to limit ourselves to describing how the process operates. As with any introductory text, we limited ourselves in the depth and breadth of material we covered. There is much more to write on the topic, and we intend to do so.

Should you have any questions or suggestions, you should feel free to contact either of us. If there are topics you want us to cover, tell us so. We welcome your input. After all, this text is intended to provide you with information you can use. The only way to remain current is to receive your input.

Bennette D. Burks
BDBurks@aol.com

Mary Margaret Minnis
MMMinnis@aol.com

http://users.aol.com/bbhogarth/hogarth.html

ACKNOWLEDGEMENTS

Many people have contributed their talents, knowledge, and skills to bring this project to completion. We would like to thank them for being responsive to our many questions and infinitely patient as we proceeded:

Jennifer Duncan and Anish Jantrania, Ph.D., of National Small Flows Clearinghouse, Morgantown, West Virginia; Alfred E. Kraus and Glynis Zywicki from Region V, United States Environmental Protection Agency, Chicago; Paul Ashburn of Ashco; Jim Baker and Bob Warfel of MultiFlo Systems; Harold and Jeff Ball of Orenco Systems; Bill Steuth of Steuth and Company; E. Jerry Tyler, Ph.D., of the University of Wisconsin-Madison, Department of Soil Science; Rein Laak, Ph.D., of RUCK Systems; Bill Hall of Kaiser-Battistone; Roman Kaminsky from Wisconsin Deptartment of Industry, Labor and Human Relations, Stevens Point, Wisconsin; Dan DeLuca, Alan Brown, Ph.D, Charlene Hoegler, Ph.D., and Ramon Lane of Pace University, Pleasantville, New York; Jeanette Semon of the Stamford, Connecticut Public Works Department; Herman Heideklang, Ph.D. and Everett Drugge of the Darien, Connecticut Environmental Protection Commission; Doug DiVesta of Stearns and Wheeler.

The chemistry and soil graphics were drawn by Thomas Janke. The microbiology figures were sketched by Kristine and Karen Hoegler. Engineering drawings were executed by Andy Hopfensperger. Jane Hogarth Minnis did the cover layout. Many of the photographs in Chapter 2 were provided by the Hach Company. Munsell contributed the color wheel in Chapter 4. The New England Interstate Water Pollution Control Commission provided the photographs depicting the pumping of a septic system. The Southeastern Wisconsin Regional Planning Commission provided the aerial photographs in Chapter 12.

The woodcut on the front cover was reproduced, with permission, from *Flushed with Pride*, by Wallace Reyburn, Pavilion Books, London.

Thanks to you and all others who have helped us along the way.

About the Authors:

Bennette D. Burks, P.E., is Director of Engineering for Consolidated Treatment Systems, Inc. in Franklin, Ohio. Previously, he was in charge of the Wisconsin Onsite Wastewater Treatment Program, where his responsibilities included research into design alternatives, development and implementation of the state's septic system administrative rules, supervision of plan review and inspection programs, and uniform enforcement by county sanitarians. He also managed a program to provide grants for the replacement of failed septic systems.

He received his B.S. and M.S. in Civil Engineering from the University of Arkansas. For his master's thesis, he developed computer software to aid in the design of wastewater treatment unit processes. In addition, holds a B.A. degree from the University of New Orleans.

Mary Margaret Minnis, Ph.D., teaches undergraduate and graduate courses in environmental science and chemistry at Pace University in Pleasantville, New York. From 1985 to 1994, she was a member of the Darien, Connecticut Environmental Protection Commission, an agency that administers wetland and watercourse regulations.

She received a B.S. degree from Marywood College in Scranton, Pennsylvania and her M.S. and Ph.D. degrees from the State University of New York - College of Environmental Science and Forestry in Syracuse, New York. Her master's thesis involved evaluating methods of purifying landfill leachate and her doctoral research resulted in a method to assign "signatures" to the particulate material that escaped from the stacks of coal-fired power plants, using a combination of scanning electron microscopy, energy-dispersive x-ray analysis and computer-based image analysis.

CHAPTER 1

INTRODUCTION TO ONSITE WASTEWATER TREATMENT

Topics of This Chapter:

- *History of Waste Disposal*
- *Indoor Plumbing's Effects Upon Waste Disposal*
- *Early Onsite Wastewater Treatment Systems*
- *Septic System Additives*

HISTORY

Humans have been concerned with proper sanitation since they first organized into tribes. Sanitation continues to be a major focus of all development. Until recently, wastewater sanitation focused on minimizing health risks, primarily infectious diseases. More recently, the scope of wastewater management issues has broadened to include chronic health risks and environmental concerns.

Systems to drain water away from buildings existed in India, Pakistan and on the Island of Crete about 2000 BC. The written western tradition of sanitation begins with the fifth book of the Old Testament, Deuteronomy, where Moses lays out the Deuteronomic Code, 23:12-14:

> *You must have a latrine outside the camp, and go out to this; and you must have a mattock*[a] *among your equipment, and with this mattock, when you go outside to ease your self, you must dig a hole and cover your excrement.*

The Romans employed water-carrying devices to send most of their wastes to nearby the River Tiber via open sewers as early as the 6th Century BC. By the 3rd Century BC, the sewers in Rome were vaulted underground networks called the Cloaca Maxima[1]. Some of these early systems even included primitive public toilets with vents, the used water from the public baths flushing away the

---

[a] A mattock is a digging tool similar to a hoe. Its blade is at a right angle to its handle.

wastes. The Romans installed plumbing devices in conquered countries, which are still in existence in Bath, England, where extensive plumbing systems carry water to and from the famous public baths. Further developments in plumbing stopped after the fall of the Roman Empire. During the Dark Ages, religious interpretations of the Bible's Deuteronomic Code precluded the use of water to carry waste. Also, the ages-old practice of separating drinking water and human wastes was largely abandoned, and human wastes could easily migrate from waste pits into wells. Epidemics raged in the cities, but the relationship between excrement and disease was not recognized.

Starting in the Middle Ages, city dwellers simply threw their wastes into the street, as shown in Figure 1.1. In some of the walled towns, organized attempts at human waste removal were started in order to clean out the privies that served several families. However, most people preferred to live near a watercourse,

Figure 1.1 - "Night" from "Times of the Day" by William Hogarth, March 25, 1736

so that the *garderobe*[b] could be positioned directly over running waters. Even the castles of Medieval England had crude plumbing that conveyed human wastes to the moat.

Throughout the Renaissance, open sewers in cities carried wastes slowly toward a watercourse, the flow augmented by the occasional rain. This was an obvious health nuisance, but no services were organized to handle the wastes. The *cesspool*, a simple pit that allowed solids to settle out and the clarified liquid to seep into the ground, made its appearance during the Renaissance. These pits were periodically cleaned out, and the contents were spread on agricultural land. But cesspools often overflowed, causing health hazards. Most rural houses continued to use privies for their wastes because no other options were available.

The English Industrial Revolution brought many people into the overcrowded cities, especially London. With the expanding urban populations came sanitation problems. Although Sir John Harrington developed a flush toilet in 1596 for Queen Elizabeth I's Richmond Palace, this device was viewed only as a curiosity. To address sanitation problems in the cities, Alexander Cumming developed a water-flushed toilet in 1775, and Joseph Bramah, a cabinetmaker in England, patented a practical flush toilet in 1778. These early toilets had no device to prevent water from continually flushing the toilet. About 1872 water-efficient flush toilet became commercially available, manufactured by Thomas Crapper & Co.

John Snow, a physician in mid-1800's England, did a pioneering epidemiological study. He traced the movement of cholera from India, over the Continent, arriving in London in 1849. He traced a recurrence of the epidemic in 1854 to a public well that was being contaminated by privy vaults[2]. Public health concerns over the storage of human wastes close to dwellings gave credibility the idea of directing all water-diluted wastes to the rivers.

Water began to be used in large quantities to move the wastes from houses. The resulting sewage was diluted and transportable, but the sewers were open and often ran down the streets. Eventually the sewers were diverted into stormwater sewer systems, running straight to the River Thames. But sewers only moved the issue of sanitation from the street to the river. In London, for example, the smell of the Thames so overpowered the members of Parliament, they began to press for some sort of sewage treatment. In a report to Parliament in 1852, Henry Austin, the Chief Superintending Inspector of the Board of Health in London, recommended a system that would separate the solids from the liquids. He provided sketches to illustrate the design, showing a multiple-tank system

---

[b] A garderobe is a closet, but in this case it is a privy.

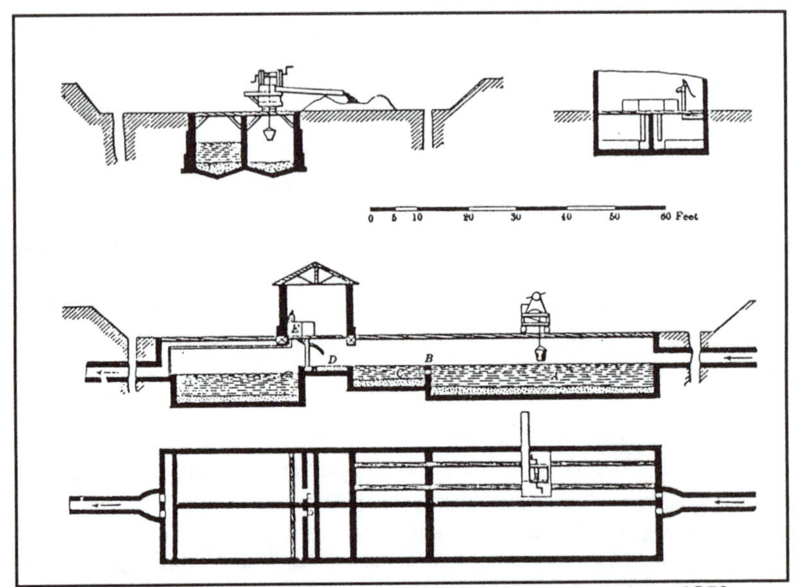

Figure 1.2 - Henry Austin's Septic Tank Design, London, 1852.

featuring coarse gravel, and the effluent was treated in a manner acceptable to the receiving body of water. Figure 1.2 shows Austin's sketches.

London was not alone in its efforts. In response to sewage disposal problems, the city of Hamburg, Germany, built an engineered sewer in 1848. In Paris, cesspools on individual properties separated the solids from the liquids; the liquids were conveyed from the city via the storm sewers. City workers periodically emptied these cesspools. Later, the law was changed to require that all sewage be discharged into the sewers for ultimate disposal on land. This system was called *tout à l'égout* or "all to the sewer."

Farmers welcomed land spreading of manure and sewage on agricultural lands. In France, England, and America, disposal of sewage on land was a common practice outside cities. In England, land disposal became a problem because the sewage overwhelmed the soil at available disposal sites, the runoff polluting nearby streams. The practice of land disposal of large quantities of municipal wastewater continued until public pressures developed for other land uses. Because of the problems with runoff, efforts were made to treat the sewage before land disposal so that the resulting liquid would be more quickly absorbed by the soil.

In 1868, Sir Edward Frankland conducted a series of experiments with London crude sewage. His device had six-foot high, ten-inch wide glass cylinders, each filled with different types of sand or soil. Through these media, he intermittently ran London's crude sewage — dosing and resting individual tanks

— and analyzed the resulting effluent. From these tests, he calculated the capabilities of different filter media to purify sewage[3]. For this reason, Frankland is considered to be the "father" of the trickling sand filter.

In 1887, the Massachusetts State Board of Health established an Experimental Station at Lawrence, Massachusetts, to experiment with innovative methods of sewage treatment. Researchers at the Lawrence facility exchanged information with counterparts in Britain and Germany. At that station in 1893, a sand bed was first used to filter the effluent from a septic tank, reducing the land areas needed for sewage disposal. The land acceptance rates were established to maintain an efficiently-working sand filter.

Hermetically-sealed tanks were not given any particular "name" during this period. The term *septic tank* was not coined until 1895, when Donald Cameron installed a water-tight covered basin to treat sewage by anaerobic[c] decomposition(Figure 1.3). He named his device the "septic tank." The following is a description of the tank, as reported by Leonard Kinnicutt:

> *The tank at Exeter, England, was an underground tank of cement concrete, 65 feet long, 19 feet wide, and with an average depth of 7 feet, and having a capacity of 53,000 gallons. The tank was covered with a concrete arch, and a portion near the inlets was made about 3' deeper than the rest and partially cut off by a low wall, forming a couple of pockets or grit chambers, to retain sand, grit and road washings. The inlet was carried down to a depth of 5' below the surface, so that air could not make its way down with the sewage, and also so that gases could not escape from the tank back into the sewer. The effluent outlet was also below the level of the liquid, and to avoid currents that might be liable to carry floating matter from the surface a cast-iron pipe was carried across the whole width of the tank 15 inches below the surface, and on the lower side of this pipe was a continuous opening about half an inch in width. An iron pipe about one and a half inches in diameter extended up out of the top of the tank to allow the escape of gases, and the whole tank could be inspected from a central manhole provided with glass window*[4].

Cameron received a U.S. Letters patent on October 3, 1899. He attempted to collect royalties from designers in the United States and brought suit against several. In response to the fury caused by Cameron's claims, Leonard Metcalf researched the history of septic tanks and, in 1901, published an article to show

---

[c] Anaerobic means "without oxygen." Anaerobic biological processes are discussed in Chapter 3.

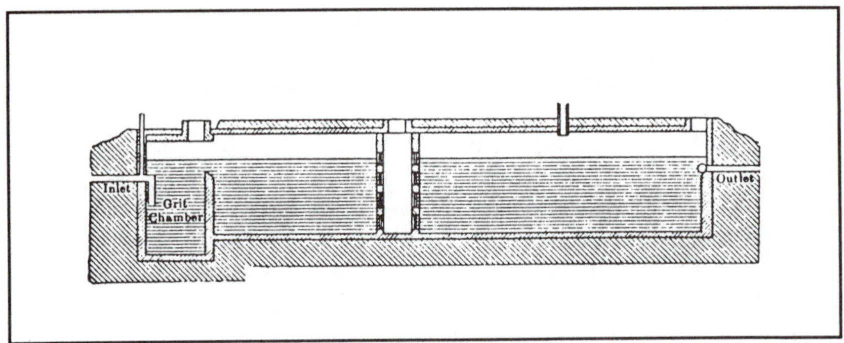

Figure 1.3 - Donald Cameron's Septic Tank, 1895

that Cameron's design was nothing new, that variations on the design had been employed in England, on the Continent, and in the United States. As a result, Cameron was unsuccessful in collecting fees, and future claims for royalties on septic tank designs were not honored. In his paper, Metcalf also described and gave drawings of many systems in the Western world. Figures 1.4*a* -1.4*d* show some of these systems, as they originally appeared in Metcalf's article[5].

- In 1858, a cemented brick pit built for a school in Derbyshire, England, received the wastes from 250 - 300 persons and a farm every day. When the pit was full, the contents were pumped to cultivated slopes. Observers said that the waters were clear and bright; they believed treatment was being accomplished with this system.

- In 1876 at the Hospital for the Insane in Worcester, Massachusetts, (Figure 1.4*a*) a settling tank to serve the waste from 600 persons was installed. It used submerged inlet and outlet pipes and the effluent was used for irrigation of the adjoining lands.

- In 1881, Abbé Moingo, editor of the *Cosmos les Mondes*, a French scientific journal, described Louis M. Mouras' *Vidange Automatique* — Mouras' Automatic Scavenger (Figure 1.4*b*). The patented design was based on experiments that involved building a tank with a glass side to observe the actions inside the tank. The system consisted of a hermetically-sealed (airtight) metal receptacle that worked continuously and was "..absolutely inodorous, which rapidly transformed all it received into a homogeneous fluid, only slightly turbid, free from any kind of deposit and odor free[6]." Abbé Moingo suggested that microscopic agents that worked only in the absence of air were responsible for the liquefaction of the solids in the tank.

The Engineering News of April 15, 1882, carried a description of Mouras' invention, so American engineers rapidly became aware of this option for onsite wastewater treatment.

- In 1882, a 42.440 L (11,000 gal) double tank system (Figure 1.4*c*) was designed at the Lawrenceville, New Jersey, School for Boys. A sub-surface irrigation system was employed for dispersal of the effluent. It treated from 23,150 to 75,700 L (6,000 to 20,000 gal) of sewage per day. The inlets and outlets to these tanks were submerged to exclude air and a vent was employed to draw off gases.

- A two-tank system was built in 1883 at the Massachusetts Reformatory, Concord, Massachusetts, to provide for 14-hour detention of night sewage. The tank was pumped daily to within 300 mm (one ft) of the bottom.

- In 1883, a large-scale test - 61,733 L (16,000 gal) per day - of the Mouras system was conducted for a facility with three toilets used by 150 people in Logelback. The inlet and outlet pipes were submerged, so the vessel was hermetically sealed. Complete decomposition of all solids took place within thirty days[7].

- In 1891, L. Paglianl, a professor of hygiene and Director of Public Heath in Rome described using Mouras Automatic Scavenger in combination with a peat[d] filter (Figure 1.4*d*) because he realized that the Parisian *tout L'égout* system was impractical in many areas where municipal sewers were inaccessible.

- Also in 1891, Leonard P. Kinnicutt et al. reported on the designs of Mr. Scott-Moncrieff, a sanitary engineer in Ashtead, England, who showed that the bacteriological purification of sewage took place in two steps and that the first step was a precursor to additional processes. Scott-Moncrieff built a small plant involving a closed, stone-filled tank. The tank (Figure 1.5) had open coke-filled trays to achieve nitrification in the second step.

Around the beginning of the twentieth century, many American and overseas communities used massive septic tanks to treat large volumes of sewage. Birmingham, England, used a septic tank that covered 2.2 ha (5.4 ac) and Saratoga, New York, boasted a 3.8 million L (1 million gal) septic system. The effluent from the septic systems was discharged directly into streams, on the land surface, and through sand filters. Land treatment of effluent was discon-

---

[d] Peat is a dark-brown or black residuum produced by the partial decomposition and disintegration of mosses, sedges, trees and other plants that grow in marshes and other wet places.

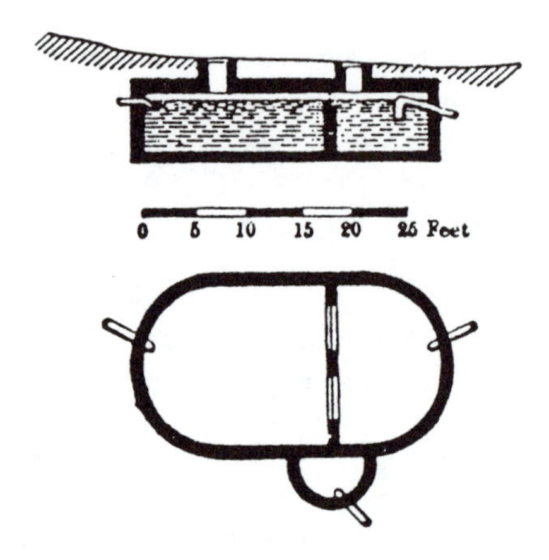

Figure 1.4*a* State Hospital for the Insane, Worcester, Massachusetts, 1876.

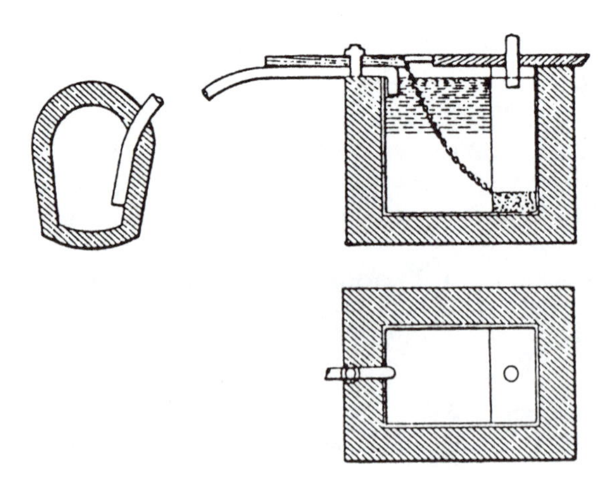

Figure 1.4*b* - Louis M. Mouras' *Automatic Scavenger*, 1891

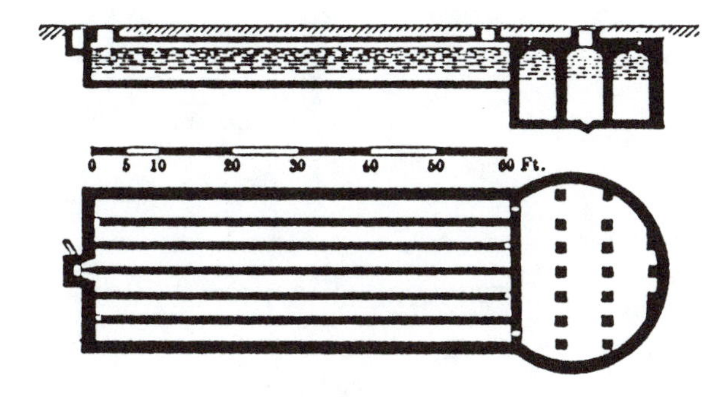

Figure 1.4*c* - Lawrenceville, New Jersey School, 1882

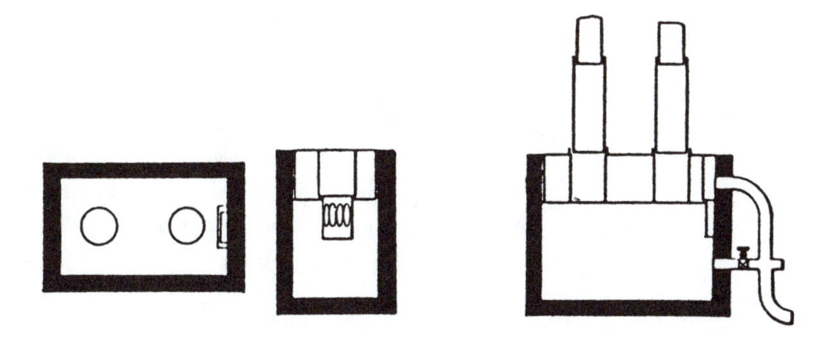

Figure 1.4*d* - Rome Public Health Director L. Paglianl's system - 1884, 1891

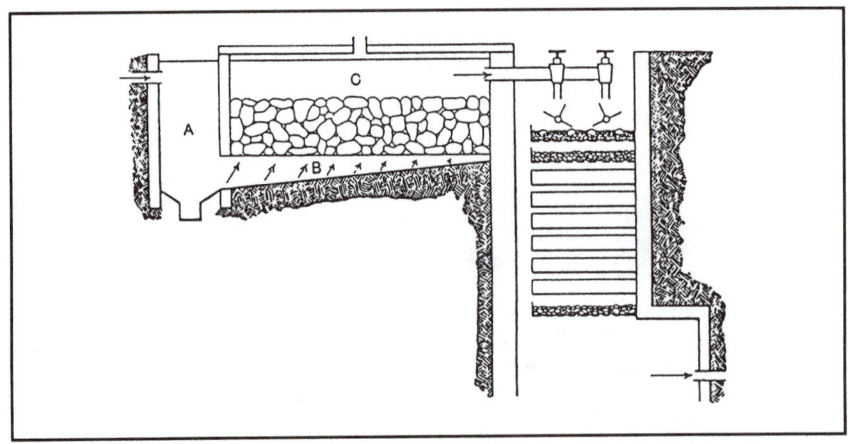

Figure 1.5 - Scott-Moncrieff's system, in 1891 in Ashted, England

tinued because insufficient agricultural land was near the treatment facilities. Sludge, the accumulation of solids from the bottom of the settling tank, was handled separately - usually by spreading it on agricultural land.

To learn more about the bacteriological action in a septic tank in 1907, Dunbar conducted a series of experiments. He set out to test the putrefaction ability of a septic tank by suspending vegetables, skinned animals, fats and other decomposable materials in an operating tank. In his report, Dunbar showed that the organic material dissolved or disappeared within three to four weeks. He showed the skeletons of the animals as proof of the putrescent ability of the septic tank[8].

ADDITIVES

Myths and legends regarding septic system additives abound. Very little factual information has been available to owners of onsite systems and, because of this dearth of reliable information, myths and legends arose. For example, a Darien, Connecticut dowager was told that she would never have a problem with her septic system if she put a dead chicken in it once a month. A California woman faithfully added one pound of ground beef to her tank every week[9]. An advertisement for one septic tank additive promises to help homeowners *avoid costly expense, avoid pumpouts, avoid wet spots, avoid re-doing the drain fields, avoid clogged drains and avoid offensive odors.*

In 1878, Alexander Mueller applied for a patent for the first septic tank additive. It was a combination of "yeasts and other fermenting substances"[10]. While yeast is still the most popular additive sold for septic tanks, there is no scientific explanation for the use of this product since yeasts prefer simple sugars as their nutrient source, not the complex protein and organic material that is in abundance in the septic tank. The continued use is probably because of septic tank folklore. The Public Health Service, in its 1949 publication *Studies on Household Sewage Disposal Systems*, showed the results of the use of yeasts and "starters" on the septic tank biodegradation process, as measured by gas production. The experiments compared the action of brewer's yeast to raw sewage, using benchtop "jar" tests. The activities of the mixtures were measured by gas production and monitored for 25 days. The mixtures with the raw sewage "starter" produced gases within two days, but the starting mixtures containing yeast did not produce any gas during the same period. For this reason, and citing two previous studies the Public Health Service determined that yeast did not effectively perform as a "starter" for a septic tank[11].

Some advertisements for products claim that it will reduce all the solids in the septic tank to gases and water. While enzymes, bacteria and fungal cultures can enhance the digestive process in the septic tank, they might interfere with the settling of solids and allow too many solids to enter the soil absorption system.

One advertisement claims that their specially-designed "superbugs" can clean grease from sewers[12]. Products sold to remove grease can actually suspend the grease in the effluent and allow that greasy effluent to move into the soil absorption system. Once the grease gets into the field, it is released from suspension and can clog the soil pores, leading to blockages of soil pores.

Onsite system owners are often confused by the advice of product manufacturers and may employ one or many products to "be on the safe side." Often, the main fear of homeowners is that their onsite system might cause an odor problem. Some support has been given to the use of Sodium Bicarbonate (baking soda) immediately before pumping the septic tank to reduce odors and make the scum more friable and easier to handle[13].

Reasons often cited for using additives include:

- reduce or eliminate the need for pumping the tank;
- increase bacterial action;
- reduce scum accumulations;
- unclog soil absorption system;
- clean the septic tank;
- deodorize the system;
- dissolve grease and other organic substances; and
- digest fats and oils.

Sometimes onsite systems actually require the addition of a substance in order to function properly. Some aerobic treatment units[e] have benefitted from specialized bacterial cultures. While specific additional substance may be required by some systems, one must learn the deficiencies or needs of that system, not just blindly add a product. For example, if a system were deficient in protein, the microorganisms would have nothing to eat and a dead chicken would provide the protein for the microbes' nutrition. That way, a more efficiently-functioning bacterial colony could be maintained[14].

The reason given most often by manufacturers of septic tank additives for improper functioning of the system is that household chemicals are highly toxic and have destroyed the bacterial colonies, so bacterial "starters" or "energizers" are needed to restore bacterial populations. The bacteria that exist in a normal septic tank are the same bacteria that are commonly found in low-oxygen soils. University of Minnesota Agricultural Extension Service states that "normal" usages of detergents, drain cleaners, toilet bowl cleaners, deodorizers and other household chemicals do not harm the septic tank bacterial populations[15].

A study of three different common household chemicals (Lysol®, Drāno® and Clorox®) was undertaken to determine the effect each of these chemicals had upon the bacteria in a septic tank[16]. Different concentrations of each product were added to the system and the effect upon the microbiological life was ascertained. In each case, the critical factor - the event that would destroy the most microbiological life - was the chemical dose. A septic system can withstand small doses of each chemical over long periods but could not withstand "slug" doses - high concentrations over short periods of time.

Based on the study, a septic tank required 9.9 L of liquid bleach, 37.8 grams of Drāno®, or 19 L of Lysol® to kill the bacteria in a 3,780 L (1,000 gal) septic tank. Recovery times needed to restore the bacterial population were: Bleach - 30 hours; Lysol® 60 hours; Drāno® - 48 hours.

Tennessee and Minnesota have issued statements to their citizens warning them about the misrepresentation of the benefits of these additives. Minnesota states that no "starter" is needed for a septic tank and that additives will only increase the amount of sludge, sometimes forcing solids out into the drainfield, possibly plugging soil pores[17]. Degreasing agents have contained toxic chemicals and carcinogenic materials that could seep into the groundwater.

Wisconsin examines additives before they are sold in the state and approves all additives just as it approves other plumbing products. But, Wisconsin does not

---

[e] Aerobic systems generally have air pumped into them to assist in the treatment of the sewage. These systems are discussed in Chapters 3 and 7.

examine the effectiveness of the additives; it simply checks to make sure the additive does not contain substances that will damage the septic system or pollute the groundwater beneath the system. Wisconsin took this step, in part, because vendors had tried to sell additives that contained volatile organic compounds.

Medications

Another set of substances that are blamed for onsite system failure are medications. Some installers cite instances when aerobic treatment units have ceased functioning; investigations into the household revealed that someone using the system had recently taken antibiotics. Aerobic systems may be more vulnerable because they are usually evenly mixed by aerators. The septic tank, partially anaerobic and facultative[f], is more quiescent and the microbial colonies are more protected against the temporary presence of antibiotics in the waste stream. Some antibiotics that are used to treat infections in the gut or urinary tract are given in gram dosages. The antibiotics are specific and either inhibit the growth of bacteria or destroy the bacteria. Human cells are quite different from bacterial cells, so the human is not affected by the antibiotic. The amount of antibiotic that is excreted is almost the amount that is ingested, so large doses of antibiotic may have profound effects on the bacteria in the onsite system[18].

Additionally, septic systems that serve nursing homes have been troubled by strange problems. It was suspected that some of the residents did not take their medications, but flushed them down the toilet.

CURRENT CHALLENGES

In the 1990's onsite wastewater treatment system designs must address four issues:

- Treatment to prevent acute health hazards.
- Treatment to reduce chronic health hazards.
- Removal of nutrients.
- Disposal of water and byproducts.

Wastewater management using onsite wastewater treatment systems ("onsite systems" or "septic systems") continue in the United States. The 1990 U.S. Census listed 24,670,875 households across the country that use onsite systems. This is 24 percent of the total households in the census. In some states, the percentage of businesses and residences served by onsite systems may surpass a third of the population. The number of onsite systems will continue to rise

---

[f]Facultative means "works in either the presence or absence of oxygen."

because of the expense and time required to construct municipal wastewater systems and suburban sprawl.

As engineers, soil scientists, and sanitarians grapple with addressing the issues listed at the beginning of this chapter, they are developing "innovations" that are sometimes over 100 years old. Such innovations that were recommended in the late 1800's include:

- Flow Reduction
- Evapotranspiration
- Shallow Soil Absorption Systems
- Lime Stabilization of Septage - the solids in the bottom of the septic tank
- Nutrient uptake by agricultural crops and shrubs
- The use of peat or carbon for final treatment
- Sand filtration of septic tank effluent

It may be that the secret of future successes lies with the wisdom of our ancestors.

REFERENCES

1. Fuhrman, Ralph E. "History of Water Pollution Control," *Journal of the Water Pollution Control Federation*, 1984, p. 306.

2. "Public Works." *New Encyclopedia Britannica.* 15th Edition. 1990.

3. Dunbar, Dr. *Principles of Sewage Treatment.* Trans. H.T. Calvert, London: Charles Griffin & Co., Ltd., 1908, p. 119.

4. Metcalf. Leonard. "Antecedents of the Septic Tank." American Society of Civil Engineers Transactions Number 909, Volume XLVI, December, 1901 (presented September 25, 1901). pp 456-481.

5. Kinnicutt, Leonard P., C.-E.A. Winslow, R.W. Pratt. *Sewage Disposal.* New York: John Wiley & Sons, 1910.

6. Metcalf. Leonard. "Antecedents of the Septic Tank." American Society of Civil Engineers Transactions Number 909, Volume XLVI, December, 1901 (presented September 25, 1901).

7. "Minutes of Proceedings" *Institute of Civil Engineers*, Vo.. LXVIII. 1883-1884: 502.

8. Dunbar, p. 92.

9. Wilcox, Kevin. "Little Common Ground in Septic Tank Additive Debate." *Small Flows*, National Small Flows Clearinghouse, Morgantown, WV. July, 1992: 1, 8-9.

10. Weibel, S.R. C.P. Straub, J.R. Thoman. *Studies on Household Sewage Disposal Systems*. Federal Security Agency, Public Health Service Environmental Health Center. Cincinnati, Ohio, 1949: 131-138.

11. Weibel. p.134.

12. "Grease Eaters Clear Sewers." Engineering News Record. McGraw-Hill, Inc. September 9, 1982.

13. Winneberger, J.H. Timothy, Weinberg, M.S. "Beneficial Effects of Baking Soda Added To Septic Tanks." Journal of Environmental Health. 38(5): 322-326, 1976.

14. Steuth, William. Steuth and Company, Washington, personal communication, September, 1993.

14. Machmeier, R.E. "Get to Know Your Septic Tank." Minnesota Agricultural Extension Service, Extension Folder 337, University of Minnesota, Duluth, Minnesota.

15. Machmeier, R.E.

16. Gross, M.A. *Assessment of the Effects of Household Chemicals upon Individual Septic Tank Performance*. Arkansas Water Resources Research Center. Publication #131, June, 1987. Fayetteville, AR 72701.

17. Machmeier, R.E.

18. Thomas R. MacGregor, Ph.D. Boehringer Ingelheim, Ridgefield, CT. Personal Communication, November, 1993.

19. Steuth, personal communication, 1993.

CHAPTER 2

WATER AND WASTEWATER CHARACTERISTICS

Topics of This Chapter:

- *The Properties of Water*
- *Characteristics of Wastewater*
- *How Nutrients Affect Waterbodies*

Water covers 71 percent of the Earth's surface, but most of it is not fit for human consumption. Only about 0.003 percent of the total is usable fresh water. That amount compares to about one teaspoon in a 55-gallon drum. Fresh water is a vital resource for countless human activities. Fresh water is found in lakes, ponds, rivers, streams, soil moisture, groundwater and atmospheric water vapor[1].

The longer water is in contact with the Earth, the more substances can become dissolved in it, making the water unfit for many uses. Nature provides a mechanism for treating water — the *hydrologic cycle* — whereby water from oceans and lakes evaporates, turns into clouds and returns to the surface as precipitation. Through this cycle, as shown in Figure 2.1, a constant supply of fresh water reaches much of the surface. When precipitation falls upon the land, the runoff takes many routes: runoff to surface water (lakes and streams), ground infiltration into aquifers, evaporation, and transpiration into the atmosphere through plant respiration.

Most people who use onsite wastewater treatment systems (onsite systems) also draw their water from *aquifers*, which are water-bearing strata of permeable rock, sand or gravel. Figure 2.2 shows how water flows in an aquifer. Many areas are completely dependent upon groundwater, and protection of aquifers is becoming more important as the population grows and puts demands upon the land located over aquifers. Therefore, it is important to protect the groundwater from contamination sources, such as insufficiently treated onsite system effluent. Surface water must also be protected because an improperly-operating onsite system near a watercourse is a source of pathogens that threaten public health. In addition, dissolved organic material in wastewater can use up the available

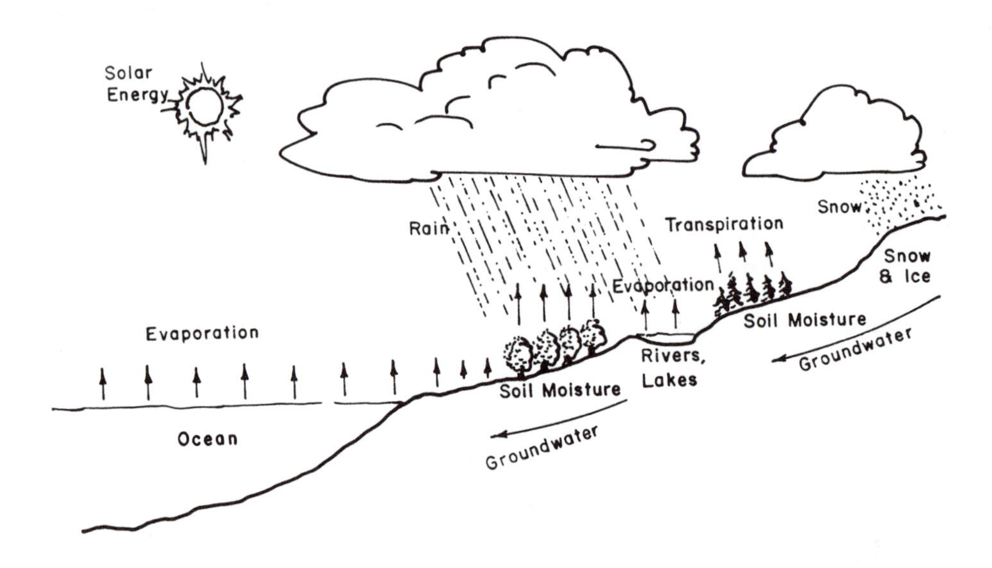

Figure 2.1 - The hydrologic cycle

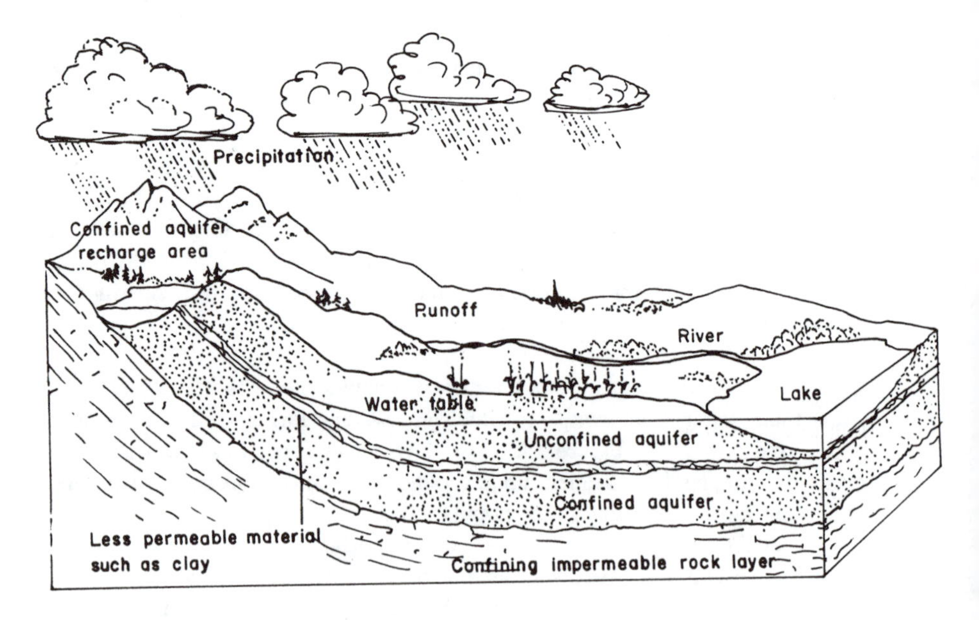

Figure 2.2 - Movement of water through an aquifer

dissolved oxygen, and nutrients can promote *eutrophication*[a] of waterbodies.

Water is a molecule unlike any other. This uniqueness is a result of the structure of the molecule and the properties of its components: one oxygen and two hydrogen atoms. Figure 2.3 shows the structure of several water molecules. Because of the nature of the oxygen atom, both hydrogen atoms are attached to the same side of the oxygen atom. Since the oxygen atom and the hydrogen atoms have different abilities to attract electrons to themselves, the hydrogen side of the molecule has a slightly positive electrical charge while the oxygen side has a slightly negative electrical charge. This results in a *polar* molecule: a molecule possessing both positive and negative poles. Because of this polarity, water molecules can easily adhere to many substances. The electrical attraction explains why so many different substances can be dissolved, suspended, or carried by water. Also, water molecules can adhere to each other. This attraction is called *hydrogen bonding* and is related to the temperature of the water; hotter water molecules do not stick to each other as much as colder water molecules.

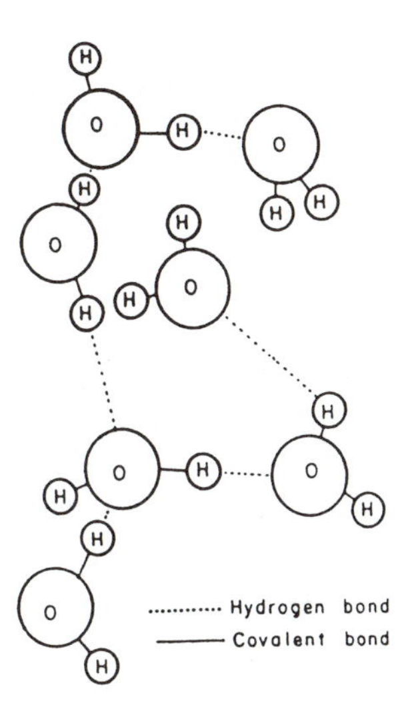

Figure 2.3
Water molecules

PROPERTIES OF WATER

Because water can dissolve, suspend, and transport so many substances, its character is highly dependent on local conditions — geology, climate, vegetation, and human activities. Its suitability as a source of drinking water, its effect on plumbing, and its ease of treatment are all greatly affected by local conditions. Many important characteristics of water that must be understood to appreciate the dynamics of an onsite system are density, alkalinity, acidity, pH, surface tension, conductivity, solubility and Biochemical Oxygen Demand (BOD).

---

[a]*Eutrophication* is the excessive growth of algae and aquatic plants in waterways. Many times, eutrophication depletes the dissolved oxygen in the water

Density

Density is a measurement of mass in relation to volume. The units used to measure density are grams per cubic centimeter (g/cm^3). By convention, water has a density of 1.000 g/cm^3 at 4°C, but the density of water varies with temperature because of the unique structural properties of the molecule. Substances that do not dissolve in water either float or sink, depending upon their densities. Substances having a density less than water will float; those having a density greater than water will sink. For example, lead has a density of 11.34 g/cm^3 and sinks in water, but gasoline has a density of 0.7 g/cm^3 and is not soluble in water, so it floats on top of water.

As water cools and the attractive forces between water molecules increase, the molecules begin to assume new positions relative to each other. This property accounts for the changing density of water. Water is at its most dense at 4° Celsius (C) — (39.2°F). When water cools below this temperature, the molecules begin to arrange themselves in a honeycomb pattern, a geometry that creates more space between the molecules. Because of this, the same mass of water has greater volume, so its density decreases. At 0°C (32°F), the molecules crystallize in this honeycomb pattern, and the resulting ice has a lower density than water — and floats.

Surface Tension

When a water molecule is far from the water's surface, its neighboring molecules exert equal forces upon it. However, the *surface* water molecules are in a unique situation: they are missing a top layer of molecules upon which they would exert an attraction. Instead, these attractive forces, which would normally be exerted upward, are discharged on the neighboring surface water molecules. Because of this increased attraction, a water surface resembles a "skin" and displays properties similar to any skin. Most liquids have surface tensions in the 20 to 40 dyne/cm range, but water is exceptional with a surface tension value of 72.75 dynes/cm at 20°C. This property has important indirect implications. For example, surface tension explains why water does not easily penetrate fabrics for cleaning. Many household cleaning products are chemicals designed to reduce surface tension. When the surface tension of water is lower, foaming, emulsification, and particle suspension may occur. These surfactant chemicals can adversely affect the operation of an onsite system by upsetting the expected process.

pH

The degree of acidity of water or wastewater is reported as *pH*. One of the most important and frequently used tests in water chemistry is pH determination.

Almost every phase of water supply and wastewater treatment is pH-dependent (for example, acid-base neutralization, water softening, precipitation, coagulation, disinfection, corrosion control). A scale, from 0 to 14, is a shorthand method for showing the degree of acidity or alkalinity of water. (Introductory chemistry texts provide a more thorough explanation.) A highly acidic solution has a pH of 1; a highly alkaline solution has a pH of 14. Pure water has a pH of 7. Every whole number on the pH scale represents a tenfold difference, so water with a pH of 4 would have ten times the acidity of water with a pH of 5 and would be one hundred times stronger than water with a pH of 6.

Figure 2.4 - Pocket-size pH meter.

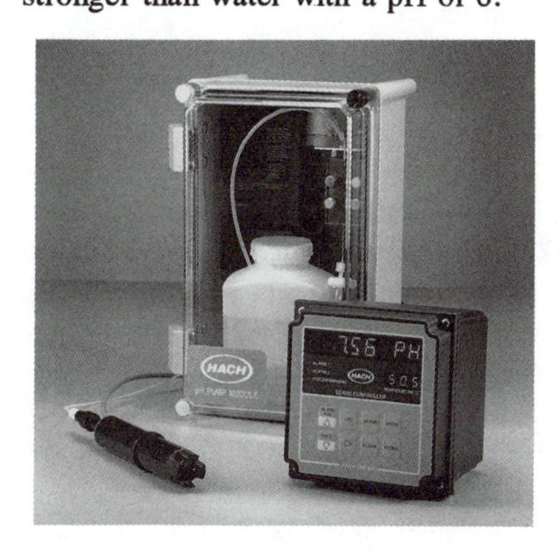

Figure 2.5 - Benchtop size pH meter

The pH of water can be found out by several methods. A rough indication of acidity or alkalinity can be made by using litmus paper, which is paper imbedded with an acid-sensitive lichen extract. Blue litmus will turn red in acidic water; red litmus will turn blue in alkaline water. For a more precise evaluation of acidity, there are pH papers made of other, more sensitive dyes that change colors within narrow pH ranges. For the most accurate determination, a *pH Meter* is used. This instrument is calibrated against a pH solution standard. There are several pH meters available for many types of needs and environmental conditions. They range from pocket size, as shown in Figure 2.4 to laboratory bench size, as shown in Figure 2.5, and should be selected according to the precision needed for the testing[2].

Solubility

Solubility, the ability to dissolve substances, is another important property of water. Water is called the *universal solvent* since it dissolves or suspends so many substances. In general, the warmer the water is, the more soluble solid substances become because the water molecules are moving more rapidly and interacting more vigorously with substances.

Contrary to this general rule of solubility, gases become less soluble in water as the temperature of the water increases. For example, at 0°C, 14 mg (milligrams) of oxygen gas can be dissolved in water. However, at 20°C, only 9 mg of oxygen can be dissolved in water[3].

The concentration of oxygen available in water is important for certain types of wastewater treatment processes. If the oxygen concentration is too low, aerobic biological reactions may not occur. Similarly, some reactions occur only when there is no free oxygen. A total lack of oxygen results in anaerobic conditions.

SOURCE WATER CONSTITUENTS

The water that is used in a structure comes either from a surface water supply, like a reservoir or river, or from a well. The source of the water determines what is dissolved in the water as it flows into the plumbing of the structure. For this reason, it is important to understand the many things that are normally present in municipal and private water supplies.

Hardness

Often, aquifers are in limestone strata, so the water drawn from them contains high concentrations of calcium and magnesium. Calcium and magnesium salts, which generally are present in water as bicarbonates or sulfates, are the usual source of water hardness. One common manifestation of water hardness is the insoluble "scum" formed by the reaction of soap with calcium or magnesium ions. Another problem caused by hard water is the formation of mineral deposits. When water containing calcium and bicarbonate ions is heated, insoluble calcium carbonate is formed and coats the surfaces of hot water systems, clogging pipes and reducing heating efficiency. The removal of water hardness is essential for many households and businesses. Often water softeners are used; they work by replacing the calcium and magnesium ions in the source water with sodium ions. When all the sodium ions on the zeolite[b] molecules are used up, the softener "backwashes" to force new sodium ions onto the zeolite. In the process, the calcium and magnesium ions liberated from the zeolite are

[b] Zeolites are large molecules, usually aluminosilicates. The zeolite anions trade their sodium ions for the calcium ions in the water.

flushed into a drain. Backwash from these systems may contain high amounts of calcium chloride ($CaCl_2$) which can seriously affect the operation of the onsite system and the capability of the soils to perform their water-purification job.

Water hardness is readily determined by EDTA[c] titration. A titration method can also be used to determine the hardness that is attributable to calcium; the magnesium hardness can then be determined by subtraction. The degree of hardness is expressed in mg/L (milligrams per liter) or ppm (parts per million) and can range from zero to hundreds of mg/L[4].

Alkalinity

Alkalinity of water is a measure of its capacity to neutralize acids or absorb hydrogen ions without significant pH change. It is the sum of all the bases that are present in the water, a combination of hydroxide, carbonate and bicarbonate ions. Carbon dioxide (CO_2), although only about 0.03 percent of the atmosphere, plays a major role in water chemistry, since it reacts readily with water, forming bicarbonate and carbonate radicals. CO_2 may be absorbed from the air or may be produced by bacterial decomposition of organic matter in the water. Once in solution, it reacts to form carbonic acid (H_2CO_3):

$$CO_2 + H_2O \rightleftarrows H_2CO_3 \rightleftarrows H^{+1} + HCO_3^{-1} \rightleftarrows 2H^{+1} + CO_3^{-2}$$

When the pH of the water is greater than 4.5, carbonic acid ionizes to form bicarbonate that, in turn, is transformed to the carbonate radical if the pH is above approximately 8.3. Carbon dioxide is very aggressive and corrodes metal water pipes.

The bicarbonate-carbonate character of water can be analyzed by slowly adding a strong acid solution to a sample of water and reading resultant changes in pH. This process of titration is used to measure the alkalinity of water. Alkalinity is measured by titrating a given sample with 0.1 N sulfuric acid. For highly alkaline samples, the first step is titrating to a pH of 8.3. The second step (or first if the water has an initial pH of less than 8.3) is titrating to an indicated pH of 4.5. The pH can be read directly from a pH meter, or colorimetric indicators (chemicals that change color at specific pH values) can be used to learn the endpoints of these titrations. Phenolphthalein turns from pink to colorless at pH 8.3; methyl orange turns from orange to pink around pH 4.5. That part of the

---

[c] EDTA is Ethylene diammine tetracetic acid. It is used in determinations of calcium concentrations in waters because it keeps calcium from chemically combining with other substances.

total alkalinity above 8.3 is called phenolphthalein alkalinity. Alkalinity is conventionally expressed in terms of mg/L as $CaCO_3$ and is calculated thus[5]:

$$\textit{Alkalinity as mg/L } CaCO_3 = \frac{(\textit{ml titrant X normality of acid X } 50{,}000)}{(\textit{ml sample})} \qquad (2.1)$$

Figure 2.6 shows a titration curve for determination of alkalinity. If the pH of a sample is less than 8.3, the alkalinity is all in the form of bicarbonate (HCO_3^{-1}) Samples containing carbonate and bicarbonate alkalinity have a pH greater than 8.3 and titration to the phenolphthalein endpoint represents half the carbonate alkalinity. If the volume of phenolphthalein titrant equals the volume of methyl orange titrant, all of the alkalinity is in the form of carbonate. Any alkalinity over this amount is due to hydroxide (OH^{-1}).

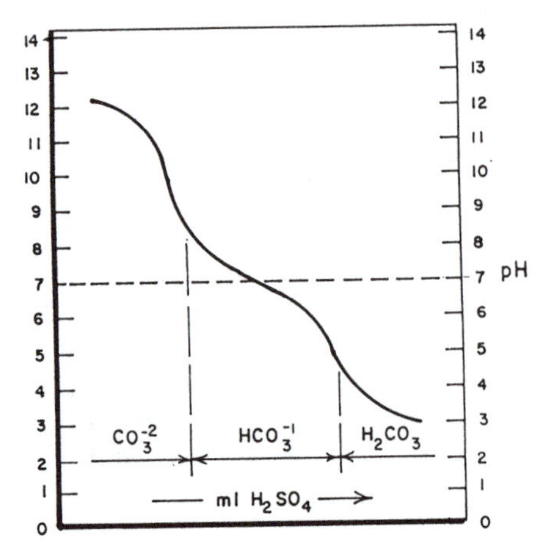

Figure 2.6 - Titration curve for alkalinity, using sulfuric acid.

In many areas of the country, the soils and mineral strata are alkaline and impart a high alkalinity to water. For example, lime ($Ca(OH)_2$) is a source of hydroxide alkalinity. Limestone ($CaCO_3$) is a source of carbonate alkalinity, and sodium bicarbonate ($NaHCO_3$) is a source of bicarbonate alkalinity. Other ions that may contribute to alkalinity are borates, phosphates and silicates. In general, the alkalinity of wastewater is only slightly different from that of the water supply.

Acidity

Acidity is the base-neutralizing capacity of water. The acidity of wastewater varies widely with the water supply and the activities of an individual household or business. An acid is any chemical compound that can produce a *hydronium ion* (H_3O^+) when it mixes with water. Examples of common acids are sulfuric acid (battery acid), hydrochloric acid (muriatic acid) and acetic acid (vinegar). Acids make water capable of dissolving many substances, especially metals. Acidic waters may be highly corrosive, influence chemical reaction rates, and affect biological processes.

Table 2.1
Composition of Typical Untreated Wastewater[6]

Constituent	Unit	Range	Typical
Total Solids	mg/L	300-1200	700
Dissolved	mg/L	250-850	500
Fixed	mg/L	150-550	150
Volatile	mg/L	100-300	150
Suspended	mg/L	100-400	220
Fixed	mg/L	30-100	70
Volatile	mg/L	70-300	150
Settleable	mg/L	50-200	100
BOD_5	mg/L	100-400	250
TOC	mg/L	100-400	250
COD	mg/L	200-1,000	500
Total Nitrogen	mg/L	15-90	40
Organic	mg/L	5-40	25
Ammonia	mg/L	10-50	25
Nitrite	mg/L	0	0
Nitrate	mg/L	0	0
Total Phosphorous	mg/L	5-20	12
Organic	mg/L	1-5	2
Inorganic	mg/L	5-15	10
Chloride	mg/L	30-85	50
Sulfate	mg/L	20-60	15
Alkalinity	mg/L	50-200	100
Grease	mg/L	50-150	100
Total Coliform	/100ml	10^6 - 10^8	10^7
VOC's	μg/L	100-400	250

WASTEWATER CONSTITUENTS

Wastewater contains a variety of substances, depending upon the activities within the structure that discharges to the onsite system. For the purposes of this text, the term "domestic wastewater" and "wastewater" will be synonymous. Both refer to wastewater generated in a "typical" house or other establishment that has similar properties. The constituents of this wastewater are listed in Table 2.1. Food service establishments have wastewater with different characteristics, so will be discussed separately. Wastewater generated from industrial facilities falls outside the scope of this text.

Biochemical Oxygen Demand (BOD)

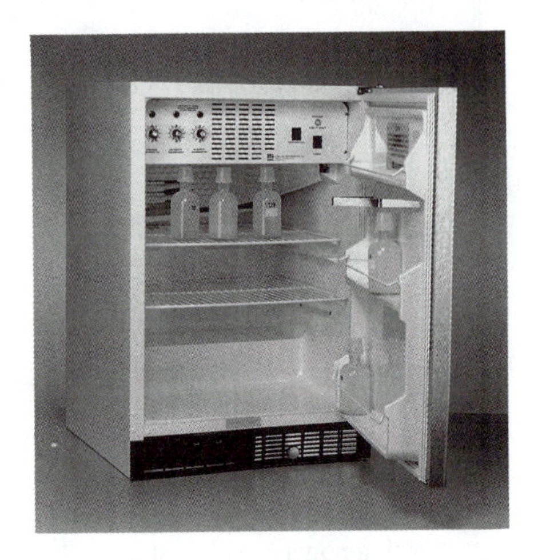

Figure 2.7 - BOD Bottles in incubator

Wastewater is composed of a variety of inorganic and organic substances contained in feces, urine, and food wastes. These substances decompose easily, but are difficult to distinguish chemically. Taken together, the substances are measured in terms of the amount of oxygen needed to support microbial consumption of the matter. This oxygen consumption is known as the *biochemical oxygen demand,* or BOD. The Five-Day BOD, or BOD_5, is measured by the quantity of oxygen consumed by microorganisms during five days (because the test originated in England where the maximum stream flow to the sea is five days).

The traditional method for determining BOD_5 consists of:

- filling a special *BOD bottle* (Figure 2.7) with diluted wastewater;
- measuring the dissolved oxygen in the sample;
- incubating the samples for 5 days in a controlled environment; and
- measuring the dissolved oxygen in the bottle after five days.

The bottles are incubated at 20 ± 1°C in the incubator as seen in Figure 2.7, in the dark, so that no photosynthesis can happen (which would produce oxygen gas). BOD is the difference between the initial and final dissolved oxygen measurement, calculating for the dilution of the initial wastewater sample.

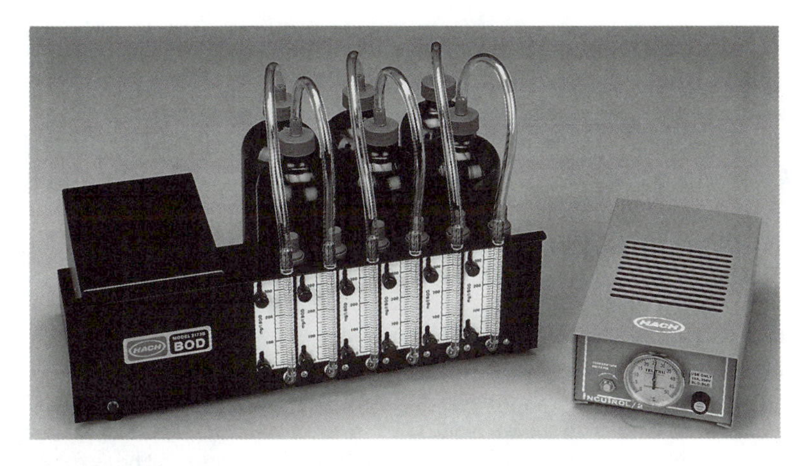

Figure 2.8 - Device to measure BOD_5 directly

BOD can also be measured directly using a manometric apparatus that measures the oxygen consumed by the sample, as shown in Figure 2.8. An electrolytic respirometer is also an instrument that measures BOD_5 directly by maintaining a constant supply of oxygen, through electrolysis of water, over the sample and monitoring the rate of generation.

Solids

Solids refer to matter suspended or dissolved in water or wastewater. Solids may adversely affect water or effluent quality in many ways. Knowing the solids concentration of the wastewater stream is helpful in determining the septic tank size and the frequency for pumping out the contents of that tank. Table 2.1 shows that the total suspended solids (TSS) in "typical" wastewater is between 300 and 1,200 mg/L. Solids analysis is important to control biological and physical wastewater treatment processes and assess compliance with effluent quality limits.

Total Solids is the term applied to the residue left in a vessel after evaporation of a water sample; it is dried to a constant weight. Total solids include *total suspended solids* -that portion of the total solids retained by a filter - and *total dissolved solids* - the portion that passes through the filter. Table 2.1 shows that the Total Suspended Solids in wastewater averages about 200 milligrams per liter. *Dissolved solids* is the portion of solids that passes through a filter of 2.0 micrometers (μm or "microns") and *suspended solids* is the portion retained on the filter.

Fixed solids is the term for the residue of total, suspended, or dissolved solids after heating to dryness for a specified time at a specified temperature and then igniting the sample at 550 $\pm$ 50°C. The weight loss on ignition is called *volatile solids*. The difference between fixed and volatile solids does not adequately differentiate between inorganic and organic matter because some inorganic minerals (such as ammonium carbonate NH_4CO_3), may volatilize[7].

Settleable solids are the materials that settle out of suspension within a designated time. A knowledge of this type of solids provides information about the amount of sludge that will be created by primary sedimentation.

It is essential to understand the distinction among the various types of solids, because each is processed differently in an onsite system. Settleable solids and some suspended solids can be removed by filtration or settling. Other suspended and dissolved solids can be treated only by biological reactions or chemical precipitation. Some dissolved solids, such as metals or chloride, may pass through an onsite system without receiving any treatment. These dissolved solids can be removed only through distillation or reverse osmosis.

Oil and Grease

Oil and grease strongly influence the operation of wastewater treatment systems. The expected oil and grease concentration must be considered during design. If present in excessive amounts, oil and grease will interfere with aerobic biological processes and lead to decreased wastewater treatment efficiency. When discharged in the effluent, oil and grease cause surface films, shoreline deposits, and environmental degradation. Grease, which is insoluble in and less dense than water, may harden in treatment tanks and can accumulate and completely clog the soil pores. Many oils stay liquid in the temperature range of an onsite system and the soil, but some solidify upon cooling. Oil or grease poured down the drain with hot water may flow through a septic tank and clog in the much cooler soil pores completely.

Oils and greases in a wastewater are not identified individually; "oil and grease" refers to all substances that are soluble in a specific solvent. Trichlorotrifluoroethane is the preferred solvent, but (due to environmental regulations concerning chlorofluorocarbons) an alternative solvent (80% *n*-hexane, 20% methyl-*tert*-butyl ether) produces similar results[8].

In typical households and small businesses, oil and grease are composed primarily of fatty matter from animal or vegetable food wastes and from petroleum hydrocarbons. A distinction must be made between these two types of oil and grease sources because of the different effects they have on the onsite system and potential environmental harm. For example, garages that have a floor drain may introduce large quantities of petroleum products into an onsite system. A knowledge of the composition of the wastewater minimizes the difficulty in determining the major source of the material, simplifies the correction of oil and grease problems in onsite systems, and prevents surface water and groundwater pollution.

For small businesses, such as restaurants and service stations, the onsite system must be designed with an integrated oil & grease separator *before* the septic tank or treatment unit. The oil and grease separator is essential to keep the onsite system from failing. The principal cause of the premature failure of an onsite system serving a restaurant is oil and grease passing through the septic tank and sealing the soil absorption system[9]. Some restaurants have separate onsite systems for kitchen and sanitary waste streams.

Table 2.1 shows that typical levels of oil and grease are between 50 and 150 mg/L. Since these substances often stay in the septic tank, domestic septage (material removed during septic tank pumping) grease concentrations between 3,850 and 9,560 mg/L have been reported[10].

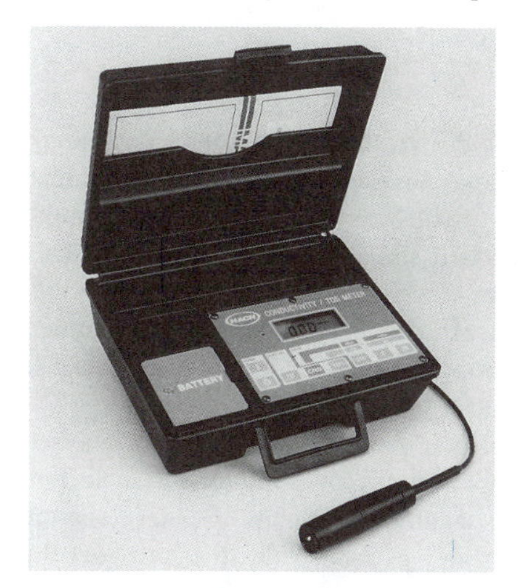

Figure 2.9 - Device to measure conductivity/TDS

Conductivity

Conductivity (or conductance) is a measure of the ability of water to carry an electric current. This ability depends on the presence of ions. The presence of ions, their total concentration, mobility, valence and the temperature of the solution all contribute to the measurement. Inorganic ions in solution provide good conductivity, but organic molecules, which dissolve but do not dissociate in solution, provide little conductivity. Conductivity is the reciprocal of resistance. Resistance is measured in ohms, and conductivity is measured in mhos, the reciprocal of ohms.

Freshly-distilled water generally has a conductivity in the range of 0.5 to 3 micromhos per centimeter (μmhos/cm). This value increases soon after the water is exposed to air in the container because of the dissolution of carbon dioxide gas (CO_2) into the sample and possible slight dissolution of the container. Potable waters in the United States range from 50 to 1,500 μmhos/cm. The conductivity of domestic wastewater may be near that of the local water supply, although the activities of the individual household or structure may cause the conductivity to vary. For instance, the backwash from a water softener, which contains high calcium chloride concentrations, will have higher wastewater conductivity.

Figure 2.9 shows an instrument to measure conductivity estimate Total Dissolved Solids (TDS). An estimate of the TDS can be derived by multiplying conductivity (in micromhos per centimeter) by an empirical factor. This factor may vary from 0.5 to 0.9, depending opon the nature of the dissolved materials. A highly saline sample will use the lower factor, while a sample low in salts but high in dissolved organic matter would require the larger factor[11]. To learn the factor, measure the conductivity of a representative sample, filter the sample, dry it dompletely and weigh the residue. This will give you the true TDS and measured conductivity from which you may determine the empirical multiplication factor. Learning about the activities in a household or business will help you to better understand the relationship between conductivity and TDS.

NUTRIENTS

Nutrients are chemicals that support plant growth. There are three *primary plant nutrients:* nitrogen, phosphorus and potassium. These three elements must be present in water-soluble forms for plants to thrive. Water-soluble forms of both nitrogen and phosphorus are abundant in wastewater and the effluent from many municipal wastewater treatment works and onsite systems. In some water-conserving places, the effluent from wastewater treatment plants is used to irrigate soils, thus treating the effluent as a resource[12].

In the United States, nutrients in wastewater from an onsite system may not e viewed as resources but as non-point sources of pollution because both surface water and groundwater can be degraded by them. Because shallow groundwater often drains into surface water as base flow, the nutrients that are present may upset the chemical balance of the receiving waters. The stepped-up addition of phosphates and nitrates to water bodies results in *cultural eutrophication*, a common problem near urban and agricultural centers. During warm weather, this nutrient overload produces dense growth of plants (algae, cyanobacteria, rooted aquatic vegetation, duckweed). The dissolved oxygen in the water body is depleted when large masses of these nuisance plants die, fall to the bottom and are decomposed by aerobic bacteria. The oxygen depletion may cause reduction in the populations of fish and other oxygen-consuming organisms. Extreme cases of cultural eutrophication can result in decomposition of organic material by anaerobic bacteria — resulting in foul odors coming from the water body. Therefore, it is important to recognize these chemicals and understand the changes they undergo in their natural cycles.

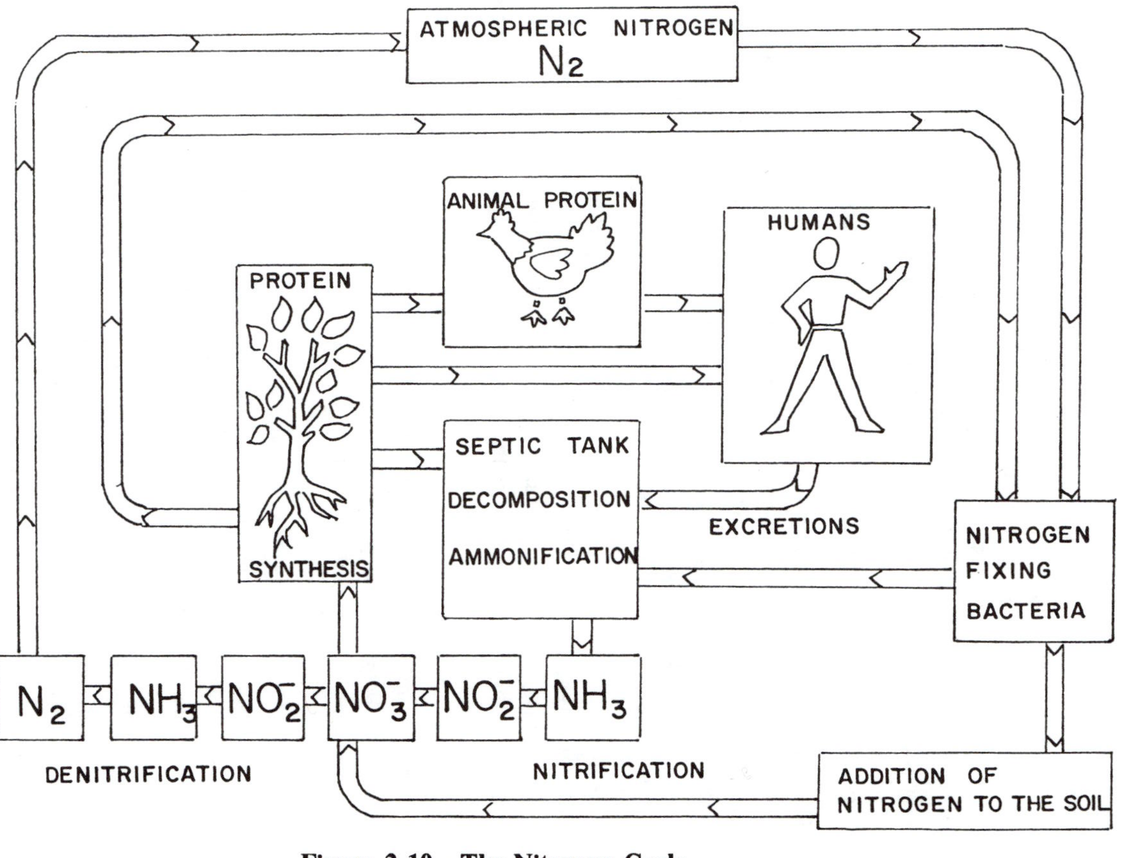

Figure 2.10 - The Nitrogen Cycle

Nitrogen

In water and wastewater, the forms of nitrogen of greatest interest are: organic nitrogen (NH_3-R, where "R" is an organic radical), nitrate (NO_3^-), nitrite (NO_2^-), ammonia (NH_4^+), and nitrogen gas (N_2). These different forms of nitrogen are all biochemically interconvertable and can be defined in terms of the *Nitrogen Cycle*. Figure 2.10 shows the various chemical phases of nitrogen as it passes through the Nitrogen Cycle. Nitrogen concentration in wastewater is often expressed in terms of the nitrogen contribution in the wastewater. For example, a concentration of 45 mg/L nitrate, expressed as nitrate, is equivalent to 10 mg/L nitrate, expressed as nitrogen. The concentrations of nitrogen compounds, expressed as nitrogen, is mg/L-N and is the unit used in this text.

Organic Nitrogen. Nitrogen that is bound to carbon includes natural materials such as proteins, peptides, nucleic acids and amino acids. Additional organic nitrogen can come from synthetic products. Organic nitrogen is a principal constituent of feces. Organic nitrogen also includes urea, H_2NCONH_2, a principal compound in urine. The normal organic nitrogen concentration of wastewater ranges from 8 mg/L-N to 35 mg/L-N.

Nitrate and Nitrite Nitrogen. These two species represent the total oxidized nitrogen. While nitrate occurs in trace quantities in surface water, nitrate may attain high levels in groundwater as a result of overfertilization or onsite systems; a limit of 10 mg/L-N is the Federal drinking water standard for total nitrogen.

Nitrate is considered the limiting nutrient (the nutrient that keeps the biotic system in balance) for primary productivity in salt waters, an important consideration in the design of onsite system along coastal areas.

The nitrate (NO_3^{-1}) ion is negatively charged, as are soil particles, so nitrate ions do not adhere to soil particles, but pass through easily. Because of public health and surface water quality issues, regulatory agencies are putting emphasis upon the reduction of nitrate concentrations in effluent from onsite systems.

Although listed separately, nitrite is not usually observed in water sources. Nitrite is readily converted to nitrate by bacterial action. Nitrite is identified because it can be observed separately from nitrate although usually in low concentrations.

Ammonium. Ammonium is usually present in surface water and is due to the chemical oxidation of urea and anaerobic processes. Unlike the mobile nitrate ion, ammonia readily adheres to the negatively-charged soil particles; therefore, ammonia is not readily leached from the soil. Rather, it is either absorbed by plant material or converted to nitrate, which is then released to the environment.

Phosphorus

Phosphorus exists in water and wastewater as the soluble orthophosphate ion (PO_4^{-3}), organically-bound phosphate (as part of large molecules in wastes), and other phosphorus/oxygen forms. Sometimes phosphate occurs in wastewater as a result of some cleaning products (automatic dishwasher detergent, for example). Most organically-bound phosphate in wastewater is from body wastes and food residue.

Phosphorus is usually the *limiting nutrient* in fresh surface water. Whenever phosphorus is the limiting nutrient in surface waters, the introduction of phosphate-containing wastewater may stimulate the growth of photosynthetic aquatic organisms and create a nuisance. In addition, as living organisms in the surface waters die and settle to the bottom, biodegradation slowly releases phosphate to the water, so a chemical reservoir of phosphate may persist long after the offending source is stopped.

The effect of phosphate is so dramatic upon some surface waters that many regulations are being considered for wastewater effluent. Often the surface of the water body is covered with a mat of algae. However, phosphate rapidly combines with other naturally-occurring chemicals, such as limestone, to form calcium phosphate. If a subsurface effluent distribution system is close to a sensitive water body, limestone added to the soil absorption system can stop the phosphate from migrating to the water body.

Table 2.2 - Effluent Quality for Class Two Streams

Parameter	Wastewater Effluent	Stream Quality
BOD_5	30 mg/L (max)	4 mg/L (max)
Suspended Solids	30 mg/L (max)	
Dissolved Oxygen		4 mg/L (min)
Total Coliforms		5,000/100ml (water supply)
		1,000/100 ml (swimming - area)
Fecal Coliforms	200/100 ml (max)	500/100 ml (water supply)

SURFACE WATER STANDARDS

Control of water quality is under Federal jurisdiction, so it is essential that standards for water quality be observed. States may have more stringent standards for the effluent that emerges from an onsite system. The quality and use of the receiving water sets the standard the onsite system must meet. Table 2.2 shows effluent and water quality requirements for the United States.

References

1. Fetter, Jr., C.W. *Applied Hydrogeology.* Charles E. Merrill Publishing Co., Columbus, Ohio. 1980. p. 4.

2. Hach Company, Loveland, CO. 1993-94 *Products for Analysis Catalog.*

3. American Public Health Association. *Standard Methods for the Examination of Water and Wastewater.* 18th Edition, Ed. by Arnold E. Greenbert, Leonare S. Clesceri and Andrew D. Eaton. 1992. 4-101.

4. American Public Health Association, p. 2-36

5. Manahan, Stanley E. *Environmental Chemistry.* Willard Grant Press, Boston, Mass. 1975. p. 264.

6. Sundstrom, Donald W. and Herbert E. Klei, *Wastewater Treatment.* Prentice-Hall, Englewood Cliffs, NJ, 1979, p. 11.

7. American Public Health Association, p. 2-55.

8. American Public Health Association, p. 5-24.

9. Steuth, William, personal communication, September, 1993.

10. Laak, Rein. *Wastewater Engineering Design for Unsewered Areas.* Technomic Publishing Company, Inc. Lancaster, PA, 1986. p. 113.

11. Walters, Gregg L. *Water Analysis Handbook* Hach company, Loveland, Colorado, p. 242.

12. Adin, A., Sachs, M. *Water Quality and Emitter Clogging Relatinships in Wastewater Irrigation.* pp. 517-530.

CHAPTER 3

MICROBIOLOGY

Topics of This Chapter

- *Types of Microorganisms Affecting Wastewater Treatment*
- *Bacterial Growth Patterns*
- *Microbiological Principles Influencing Wastewater Treatment*
- *How Microbiological Tests Ensure Safe Water Supplies*

Microbiology is the study of microorganisms and their activities. Environmental microbiology concerns itself with microorganisms commonly found in water, wastewater, and soil that may perform useful functions (decomposing organic matter) or affect public health.

BACTERIA

Bacteria are one of several important groups of microorganisms. They are essential to the nutrient cycle of the ecosystem. Many different types of bacteria are important in water and wastewater treatment processes and in the natural self-purification of streams and lakes. They are also important in the decomposition of materials in soils, in compost heaps, and in the decomposition of materials in landfills. Bacteria play several roles in recycling nutrients in the aquatic environment, as was seen in Chapter 2, *Water and Wastewater Characteristics*.

Bacteria are unicellular microscopic organisms. They are found in water, wastewater, soil, air, milk, on and in plants, animals, and many parts of the human body. Bacteria reproduce by dividing (binary fission) and are characterized by their size, shape, structure and arrangement of cells. Individual bacteria have one of three general shapes:

- spherical (coccus, cocci)
- cylindrical or rod-like (bacillus, bacilli) and
- spiral (spirillum, spirilla).

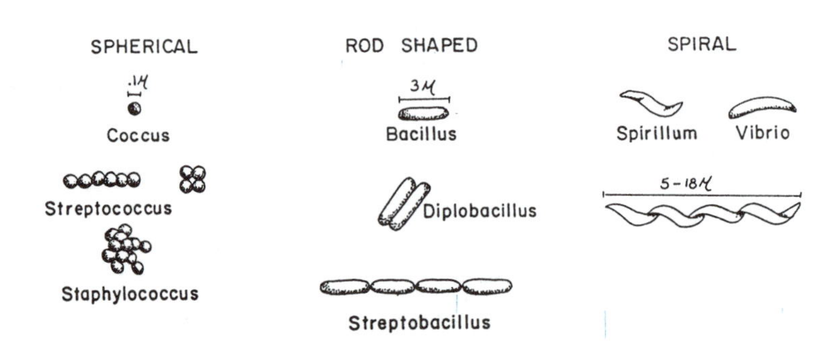

Figure 3.1 - Three general types of bacteria.

Bacterial cells may be arranged in groups such as pairs, clusters, or chains. Most bacteria range in size from 0.5—5.0 μm (microns) long and 0.3—1.5 μm wide. A single coccus is about 0.1 μm in diameter. Figure 3.1 shows the three main types of bacteria that are commonly found in natural environments.

Because bacteria are so small, they have a large surface to volume ratio, allowing them to absorb nutrients rapidly and those nutrients quickly diffuse when they get inside the cell. This characteristic enables bacteria to reproduce very rapidly. For example, the bacterium *Escherichia coli (E. Coli)* can duplicate itself in about twenty minutes in a hospitable environment.

All bacteria, as shown in Figure 3.2, have a rigid cell wall that maintains the shape of the cell and protects its contents from osmotic pressure (a flow of water molecules that tries to equalize pressure on both sides of a membrane). If that cell wall were removed, the cell might collapse or burst from the pressure of its contents. The wall is usually from 0.02 to 0.03μm thick and accounts for 10 to 40 percent of the dry weight of the organism. Some

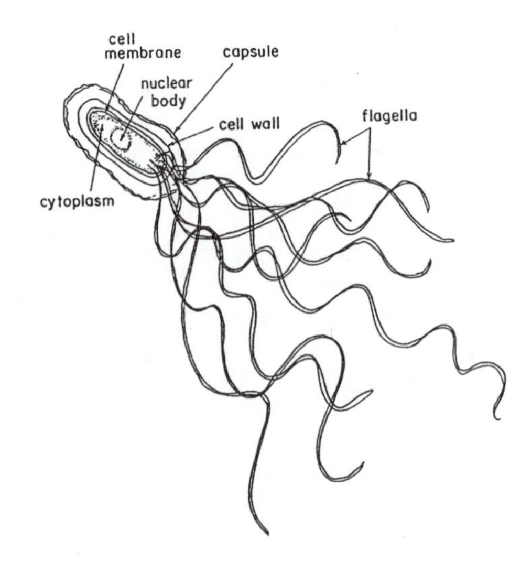

Figure 3.2 - A bacterium

antibiotics, like penicillin, work by disintegrating cell walls. Since human cells do not have cell walls, the bacteria die but the human does not.

Under the *cell wall* of the bacterium lies the *cytoplasmic membrane*, a semi-permeable boundary separating the cell's protoplasm from the external environment. This membrane allows nutrients to pass into the cell and waste products to pass out. Damage to this membrane by chemical or physical agents can cause the death of the cell.

Inside the cytoplasmic membrane is the cytoplasm, a jellylike substance containing many enzymes that can break down cellular food; Ribonucleic Acid (RNA) is suspended in this substance and gives the material a granular appearance. The *nuclear body* is suspended in the cytoplasm and contains the Deoxyribonucleic Acid (DNA) for the cell. DNA is a set of very long carbon chains, arranged in two strands called a double-helix molecule. It contains, at regular intervals on each carbon chain:

- phosphoric acid,
- a 5-carbon sugar,
- a nitrogen-containing base (adenine, guanine, cytosine or thymine).

DNA controls cell growth and reproduction and is responsible for the genetic identity and stability of the species. The basic structure of the DNA molecule is shown in Figure 3.3. This molecule is often the source of much of the nitrogen and phosphorus that is found in wastewater. Some bacteria are covered by a layer of sticky material, called the *capsule* or slime layer. It is believed that this sticky substance is excreted from the cell. This micropolysaccharide material is responsible for the infective capacity of some bacteria.

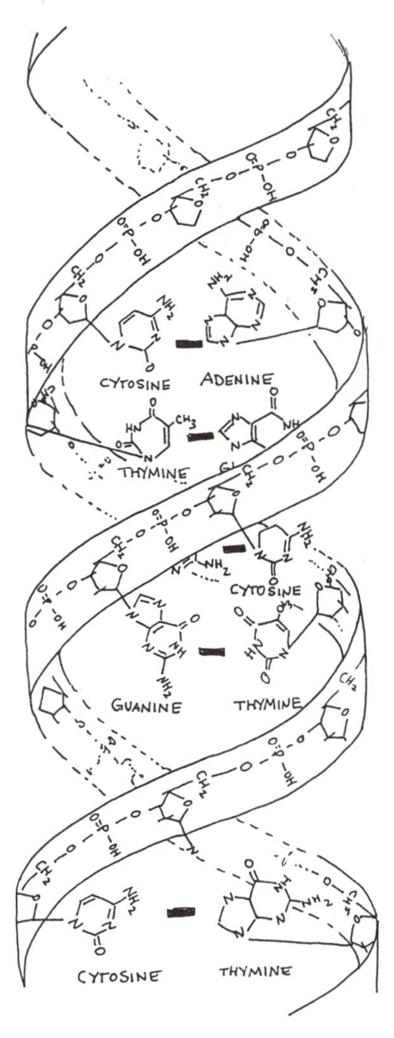

Figure 3.3 - The DNA Molecule

Most bacteria are *heterotrophs* — organisms that use carbon or dead preformed organic molecules as their energy and food source. Another type of bacteria, the *autotroph* is capable of manufacturing its own food from carbon dioxide, water and sunlight in a process called *photosynthesis*.

Many bacteria can move around rapidly in liquids by undulating or lashing one or more of their whip-like *flagella*. Flagella are extensions composed of the protein flagellin; they are found on many bacilli and spirillium types of bacteria. The flagella on the bacterium in Figure 3.2 are called polar, since they appear only on one side.

FUNGI

Fungi are a group of *heterotrophic* microorganisms that are widely distributed in nature. Fungi are considered *saprophytes* because they obtain their nutrients from the decomposition of **dead** organic matter. Fungi convert complex organic matter into simple chemical compounds that can then be used as nutrients by other organisms. The fungi characteristically form *hyphae*, elongated cells or chains of cells that absorb nutrients from their environment. These hyphae grow by extension of their tips. Most fungi can produce sexual and asexual spores[1]. The spores allow fungi to lie dormant, not reproducing, for long periods when environmental conditions are unfavorable; when conditions again favor fungal growth, the spores release the fungi to reproduce. Fungi are abundant in soils and are very

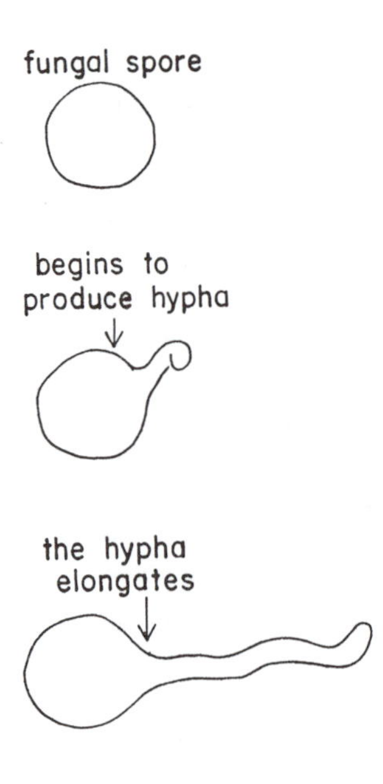

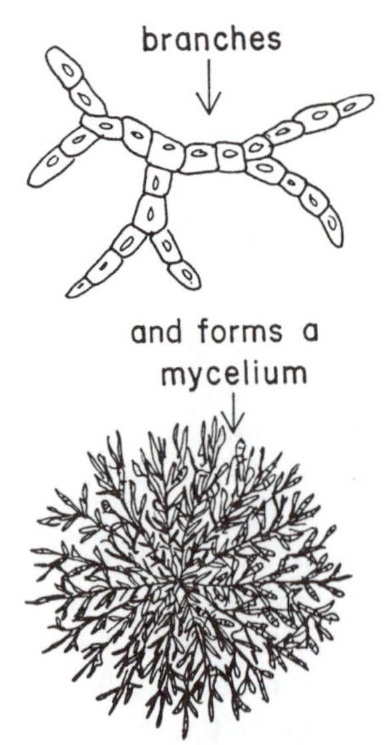

Figure 3.4 - Life Cycle of a Fungus.

important to effective biodegradation of the effluent from an onsite wastewater system. Figure 3.4 shows different stages in the life cycle of a type of fungus.

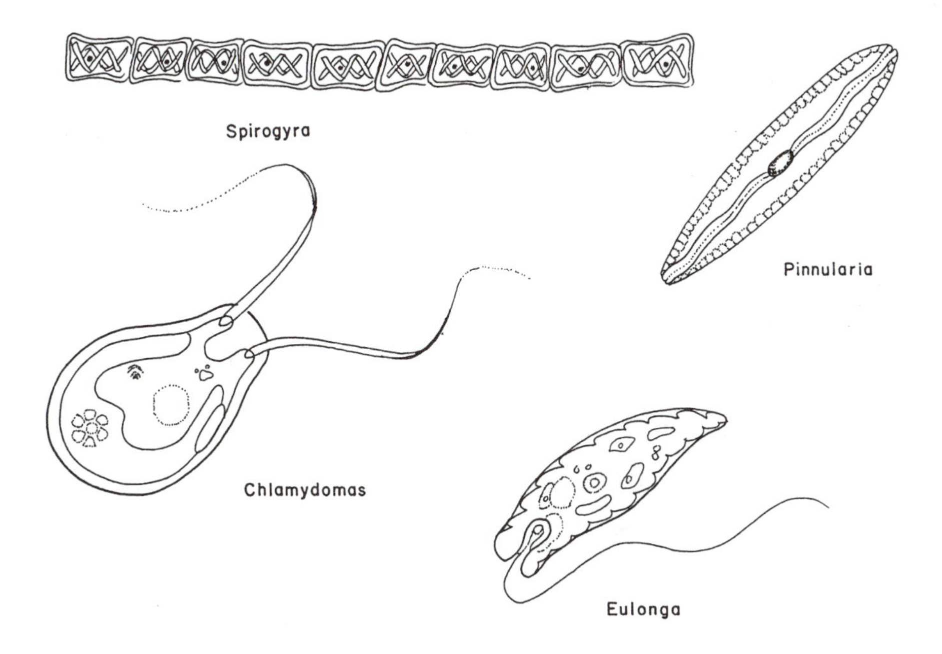

Figure 3.5 - Examples of Algae

ALGAE

Algae are usually unicellular organisms that may be nitrogen-fixing (Nitrogen fixing is the chemical conversion of atmospheric nitrogen into water-soluble nitrogen compounds.). As nitrogen-fixing photosynthetic plants, algae create large amounts of living tissue (*biomass*), which serves as food for aquatic-dwelling organisms. Additionally, photosynthesis generates molecular oxygen (O_2) as a byproduct. This oxygen, when released in aquatic environments, helps to maintain the dissolved oxygen level in natural waters. However, algae can overproduce whenever excess nutrients are provided. Frequently, the effluent from an onsite system provides phosphate, the limiting nutrient that stimulates algal blooms in fresh water lakes, and nitrate, the limiting nutrient in coastal estuaries. Figure 3.5 shows several types of algae that influence wastewater treatment processes.

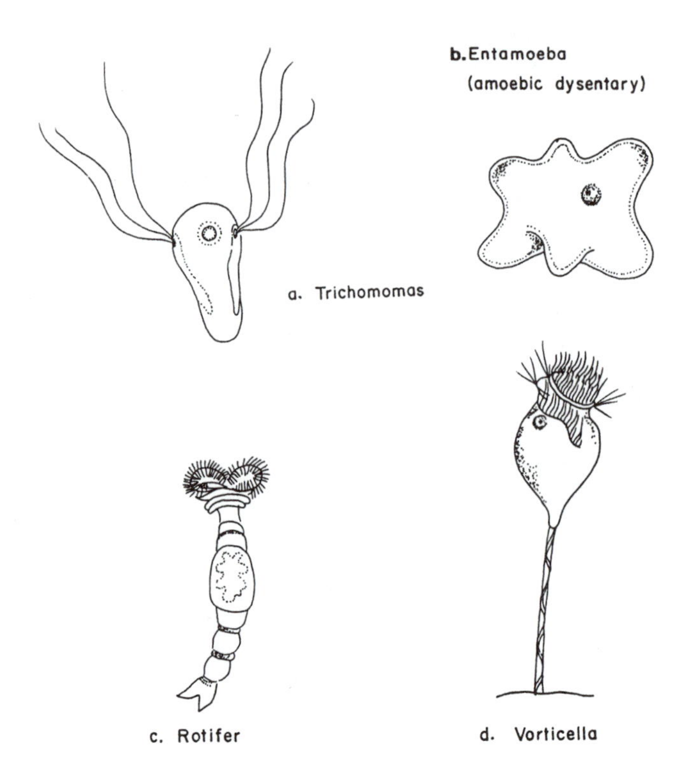

Figure 3.6 - Examples of Protozoa

PROTOZOA

Protozoa (singular - protozoan) are single-celled aquatic animals that multiply by binary fission. They have complex digestive systems and ingest solid food. Protozoa are aerobic organisms found in activated sludge, trickling filters, and oxidation ponds treating wastes, as well as in natural waters. By ingesting bacteria and algae, they provide a vital link in the aquatic food chain.

Flagellated protozoa are the smallest type, ranging in size from 10μm to 50 μm. Figure 3.6 shows protozoa that are important to the ecology of a wastewater treatment system. Long hairlike strands or *flagella*, provide motility by a whiplike action, as shown in *Trichomonas* (Figure 3.6a). Many protozoa ingest solid food or soluble organic material. *Amoeba* (Figure 3.6b) are commonly found in the slime coating on trickling filter rocks and aeration basin walls. *Rotifers* (Figure 3.6c) are simple, multicellular, aerobic animals that metabolize solid food. Rotifers are found in natural waters, stabilization ponds and extended aeration basins under low organic loading. *Vorticella* (3.6d) are the most frequently-appearing attached ciliates and are usually associated with high organic loading and low oxygen supply[2]. Many of these larger types of organisms can be recognized under a field microscope[3]. During a microbiological examination of an activated sludge sample, the protozoa stand out much more noticeably than the bacteria, but their relative influences on the treatment process compared with the bacteria is comparatively slight. As they principally feed on bacteria and very fine suspended particulate matter, protozoa contribute chiefly to a reduction of the turbidity of the treated effluent[4].

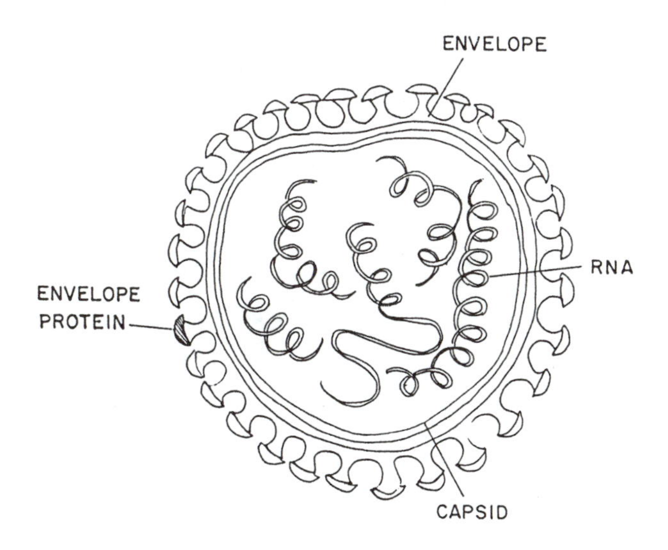

Figure 3.7 - One type of Virus

VIRUSES

Viruses are unique in that they contain no internal enzymes and cannot grow or metabolize on their own. They are *obligate parasites*, infecting the tissues of bacteria, plants and animals — including humans. Obligate parasites are organisms that can reproduce only if they get their genetic "tail" into a living host cell. An individual virus is only 1/100th the size of the smallest bacterium, usually from 10 nm to 250 nm (nanometers, 1nm = 1×10^{-10} meters). According to cell theory, viruses are on the borderline of life — they contain either DNA or RNA as their core protein — not both. Because they are not really alive outside a host cell, some viruses survive for a long time (sometimes hundreds of years) between infections and can only be killed by alteration of their molecular structure. However, other viruses, like the AIDS virus, die very quickly outside the body[5]. Figure 3.7 shows one type of virus shape.

There are more than one hundred distinct types of human enteric (intestinal) viruses, some more pathogenic than others. The usual transmission route for enteric viral infections is through the fecal/oral route, but transmission has been known to happen via water supplies, so it is important that an onsite system reduce viruses as the effluent percolates through the soil. Enteric viruses multiply only within living, susceptible cells, so they cannot multiply in sewage.

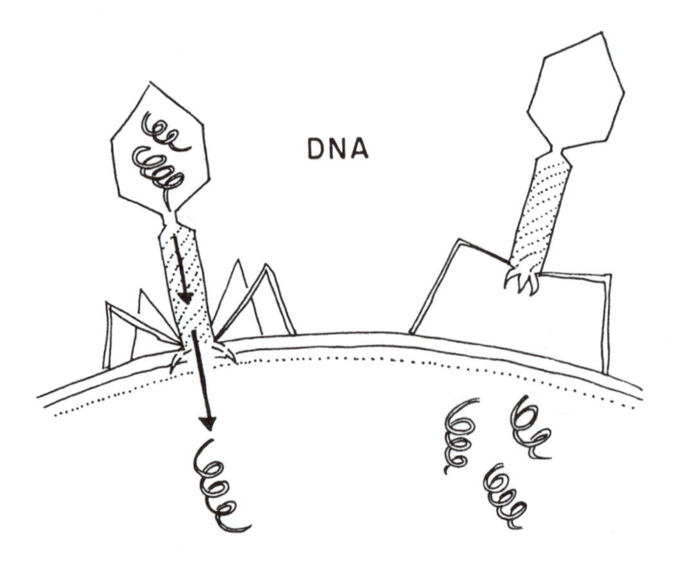

Figure 3.8 - Viral Infection Process

Different types of soil have the capability to adsorb viruses to themselves, thus reducing the virus concentration of the effluent from a septic system. However, individual virus types have different abilities to affect humans; sometimes a very small concentration of a virus in a water supply can cause a human infection. Not all viruses cause human diseases and some can even be beneficial. Examples of pathogenic viruses are those that cause smallpox, infectious hepatitis, influenza, and polio. Figure 3.8 shows how a viral infection can take place.

BACTERIAL LIFE CYCLES

All living organisms, including bacteria, have nutritional and physical requirements that must be met to sustain life. Among the species of bacteria, there are wide variations in nutritional requirements and the physical conditions they can withstand. Certain bacteria grow at temperatures below 0°C, others at temperatures as high as 99°C. Some bacteria require atmospheric oxygen; others are hindered — or killed — by its presence.

Besides requiring carbon for life and growth, heterotrophic bacteria require other nutrients, which include nitrogen, sulfur, phosphorus, and traces of metals such as magnesium, calcium, and iron. Some bacteria are more adaptable than others. For example, *E. coli* can manufacture its own vitamins from other compounds. Other bacteria are fastidious and grow only under specific conditions and in the presence of an array of specific nutrients.

Gases

Oxygen is the most important gas involved directly in microbial growth. Because oxygen is important, bacteria are divided into the following groups based on their need for oxygen:

- *Aerobic* bacteria require free oxygen for growth.
- *Anaerobic* bacteria can grow without free oxygen.
- *Facultative* bacteria can grow with or without oxygen.
- *Microaerophilic* bacteria grow in the presence of minute quantities of free oxygen.
- *Anoxic* bacteria grow in the presence of "combined" oxygen, such as the nitrate ion, NO_3^-.

The adjectives *facultative* and *obligate* describe the degree of dependence on a particular condition. For example, an obligate anaerobe is a bacterium that will not grow in the presence of free oxygen. A facultative autotroph is an organism that uses CO_2 as a source of carbon but can also grow heterotrophically with organic compounds as energy sources.

Another factor influencing bacterial growth is pH. Most bacteria exhibit optimum growth at a pH range from 6.5 to 7.5, with a maximum range for growth between pH 4.0 and 10.0[5]. The metabolic activities of bacteria can cause shifts in the pH of their environment. To avoid a kill-off of bacteria, the environment must have a *buffering capacity,* which is the ability of aquatic systems to withstand the addition of either strong acids or bases without significant changes in pH (See Chapter 2, Water and Wastewater Characteristics, for details). The pH must remain unchanged for bacterial growth to continue for an extended period.

Other physical conditions that are important for some species of bacteria are, for example, salt concentration and the presence of moisture to carry the necessary dissolved nutrients.

Growth and reproduction of bacteria

Growth and reproduction of a cell occurs as nutrients are taken into the cell and transformed into new cell components. The most common bacterial reproductive process is called *binary fission*, a characteristic of bacterial growth. DNA is replicated and distributed in the cell. A cell wall or *septum* develops which divides the bacteria and separates it into two living cells.

Bacterial populations can reach high densities very quickly. The individual cells double at a rate characteristic for each organism. This interval is known as the *generation time.* Generation times at 20°C range from 15 to 20 minutes for *E. coli* to several hours for other *mycobacterium* species. Although bacteria can grow in many conditions, optimum growth rates are dependent on specific environmental conditions for each species.

The rate of growth of a bacterial population is directly proportional to the number of bacteria present. Therefore, growth rate is a first-order kinetic process. This can be expressed mathematically as:

$$\frac{dB}{dt} = cB \tag{3.1}$$

where

$\frac{dB}{dt}$ = k, Rate of growth
B = Bacteria concentration at time t
c = First-order growth-rate constant

The equation can then be simplified to:

$$\left(\frac{1}{B}\right) dB = c\,dt \tag{3.2}$$

Integrating each side yields:

$$\ln B = ct + d \tag{3.3}$$

From this, it follows that:

$$B = e^{ct+d} = e^{ct} e^{d} \tag{3.4}$$

If B_o is the concentration of bacteria at time = 0, substituting into the last equation yields the following:

$$B_o = e^{d} e^{0} = e^{d} \tag{3.5}$$

From this, bacterial growth can be calculated thus:

$$B = B_o e^{ct} \tag{3.6}$$

This type of logarithmic growth is typical of only a small portion of the normal growth pattern of a microbial population. Figure 3.9 shows a typical growth curve, showing an initial period of little or no growth called the *lag phase*, a period of rapid growth called the *log* or *exponential growth phase*, and a period of stabilization called the *stationary phase*. As wastes accumulate and nutrients and gases are utilized, the number of cells dying outnumber those being reproduced and the *death* or *endogenous* phase occurs.

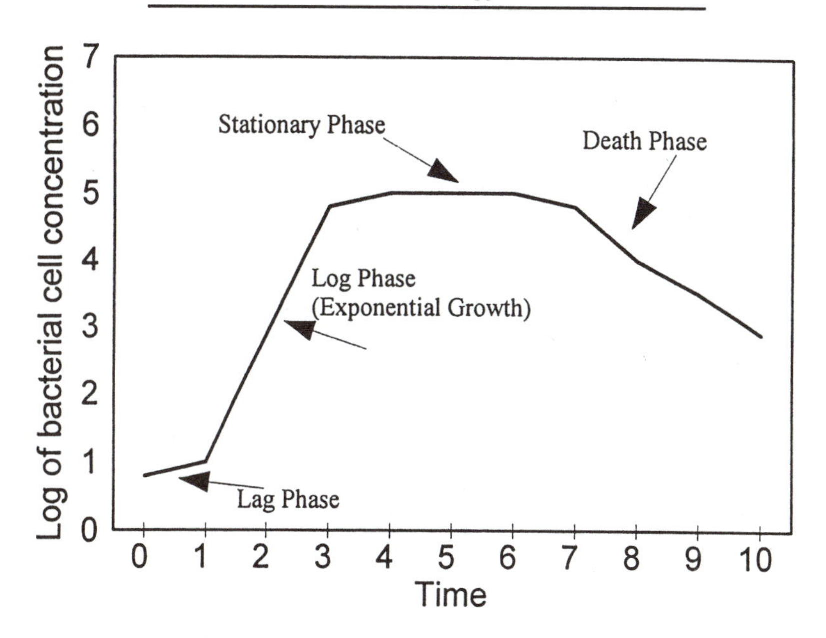

Figure 3.9 - Typical Bacterial Growth Pattern

During the initial period or *lag phase*, the cells adjust to their new environment. They may be deficient in certain growth factors and may have to adapt to other conditions. After adaptation, the cells divide and start reproducing in the *log* or *exponential growth* phase, during which time the population doubles at regular intervals. It is the phase of greatest growth under optimal conditions. The microbial population is most uniform in terms of chemical composition, metabolic rates, and other physiological characteristic. Growth cannot continue indefinitely because of food limitations, so the cells start to die. An increasing death rate depresses the net population growth until zero population growth results.

When the number of new cells being produced equals the number of cells dying, a dynamic equilibrium is reached at which there is no further population increase. This is the *stationary phase*. The reason for cessation of the growth is usually due to the exhaustion of critical nutrients.

The *endogenous phase* is reached when the death rate exceeds the growth rate. This condition results when either nutrients are depleted or toxic by-products of cell metabolism increase in the environment, inhibiting growth further.

In continuous biological wastewater treatment processes, the bacterial population degrading the waste organic matter is maintained in the stationary phase by the removal of excess growth. Some wastewater treatment processes use the endogenous phase where cells use protoplasm from dead cells to obtain energy and thus reduce the sludge volume.

In aerobic processes, heterotrophic bacteria oxidize about one-third of the colloidal and dissolved organic matter to stable products (CO_2 and H_2O) and convert the remaining two-thirds into new microbial cells that can be removed from the wastewater by settling. The complete biological conversion proceeds sequentially, with oxidation of carbonaceous material as the first step, i.e.,

$$\text{Organic matter} + O_2 \longrightarrow CO_2 + H_2O + \text{new cells}$$

Under continuing aerobic conditions, autotrophic bacteria (those obtaining carbon from inorganic compounds) then convert the nitrogen in organic compounds to nitrate according to the simplified equations

$$\text{Organic N} \longrightarrow NH_3 \text{ (decomposition)}$$

and

$$NH_3 + O_2 \xrightarrow{\text{nitrifying bacteria}} (NO_2^-) \longrightarrow NO_3^- \text{(nitrification)}$$

The conversion of ammonia to nitrite is shown as an intermediate step. In the environment, nitrite is quickly oxidized into nitrate. No further changes in the nitrate takes place unless the process becomes anoxic. Under such conditions, heterotrophic bacteria convert the nitrate to odorless nitrogen gas:

$$NO_3^- \xrightarrow{\text{denitrifying bacteria}} (NO_2^-) \longrightarrow N_2$$

These naturally-occurring reactions are used in various biological processes for the treatment of wastewater. In every case, nutrients essential for biological growth must be present in the waste or must be added to the wastewater. For municipal wastewater treatment, aerobic processes have predominated because of their simplicity, stability, efficient and rapid conversion of organic contaminants to microbial cells, and relatively odor-free operation.

Anaerobic Processes

In anaerobic biological processes, two groups of heterotrophic bacteria, in a three-step liquefaction/gasification process, convert much of the organic matter initially present to intermediates, principally organic acids, then to methane and carbon dioxide gases.

The first step, called *acid fermentation* or *hydrolysis*, involves the decomposition of organic material to complex organic acids. In the second step, called *acid regression* or *acidogenesis*, the products of the first step are decomposed into simpler products that can be eventually converted, during the third step, into methane and carbon dioxide. The third step, called *alkaline fermentation* or *methanogenesis*, is the actual release of energy by the conversion of the methanogenic substrates to methane and carbon dioxide.

The methane bacteria are some of the oldest forms of life on Earth, part of the group called *Archaebacteria*. Since they became segregated from other forms of life early in evolution, they have distinctive features, including a unique type of cell wall[6].

Anaerobic digestion is usually done in a heated anaerobic digester, where primary and secondary sludges are retained for approximately 15 to 30 days at 35°C to reduce the total solids about 40 percent and thus simplify disposal (usually on agricultural ground).

Two major advantages of anaerobic processes over aerobic processes are:

- Production of energy as methane; and
- Sludge production only about 10 percent compared to aerobic processes.

A disadvantage is that heat is needed and the reaction rate for anaerobic processes are slower than aerobic processes.

Wastewater Treatment Biology

Figure 3.10 shows several microorganisms that are normally found in aerobic wastewater processes. *Ciliated protozoa* are essential for the conventional activated sludge process, and the quality or condition of a sludge can be diagnosed by the types of microorganisms present. In addition, each microorganism has a specific growth curve and responds to different environmental conditions, including pH, temperature, nutrient concentration, oxygen availability, and mixing[7]. A "satisfactory" sludge has a preponderance of ciliated protozoa, which move and feed with the aid of cilia (fine hair-like processes). Among the ciliates found in a suitable sludge are: *Carchesium, Vorticella, Euplotes, Epistylis, Lixiogtkkynm Choenia, Chilodon, Colpoda, Colpidium, Paramecium* and *Aspidisca*[8]. Very few flagellate protozoa, amoebae and rhizopoda are present in a "good" sludge, and filamentous growths are either absent or present in only small amounts. Rotifers are often present. Unsatisfactory sludge contains predominantly flagellate protozoa that have one or more flagella (fine, whip-like processes) while few ciliates are present.

Filamentous microorganisms are those that grow in a threadlike form and under adverse conditions. When *fungi* or species such as *actinomycete* grow in a reactor, they trap water in their network of threads and prevent the sludge from

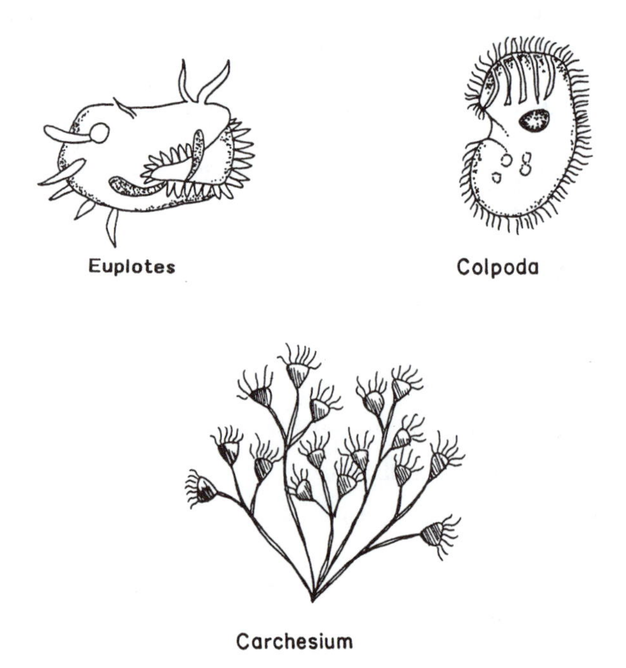

Figure 3.10 - Organisms typically found in a Aerobic Systems

settling properly. Another type of filamentous microorganism, of the *nocardia* species, can cause a viscous, brown foam with an obnoxious odor in the aeration basins. Conditions in a reactor can be controlled to avoid the growth of filamentous microorganisms. The conditions under which filamentous microorganisms develop are: fluctuations in strength, nature and flow of the wastewater, pH levels, temperature, nutrient content, dissolved oxygen levels, low food to microorganism ratios and the presence of toxins.

Soil Microbiology

Most land-based living things — plants, animals and protista — and their associated wastes eventually find their way into the soil. There, microbial activity transforms this material into the substances that constitute soil. Without this activity, nutrient cycles such as the nitrogen cycle would not be complete, and life on earth would be affected.

Bacteria and fungi constitute the largest group of microorganisms in soils. Autotrophic and heterotrophic bacteria degrade complex organic and inorganic

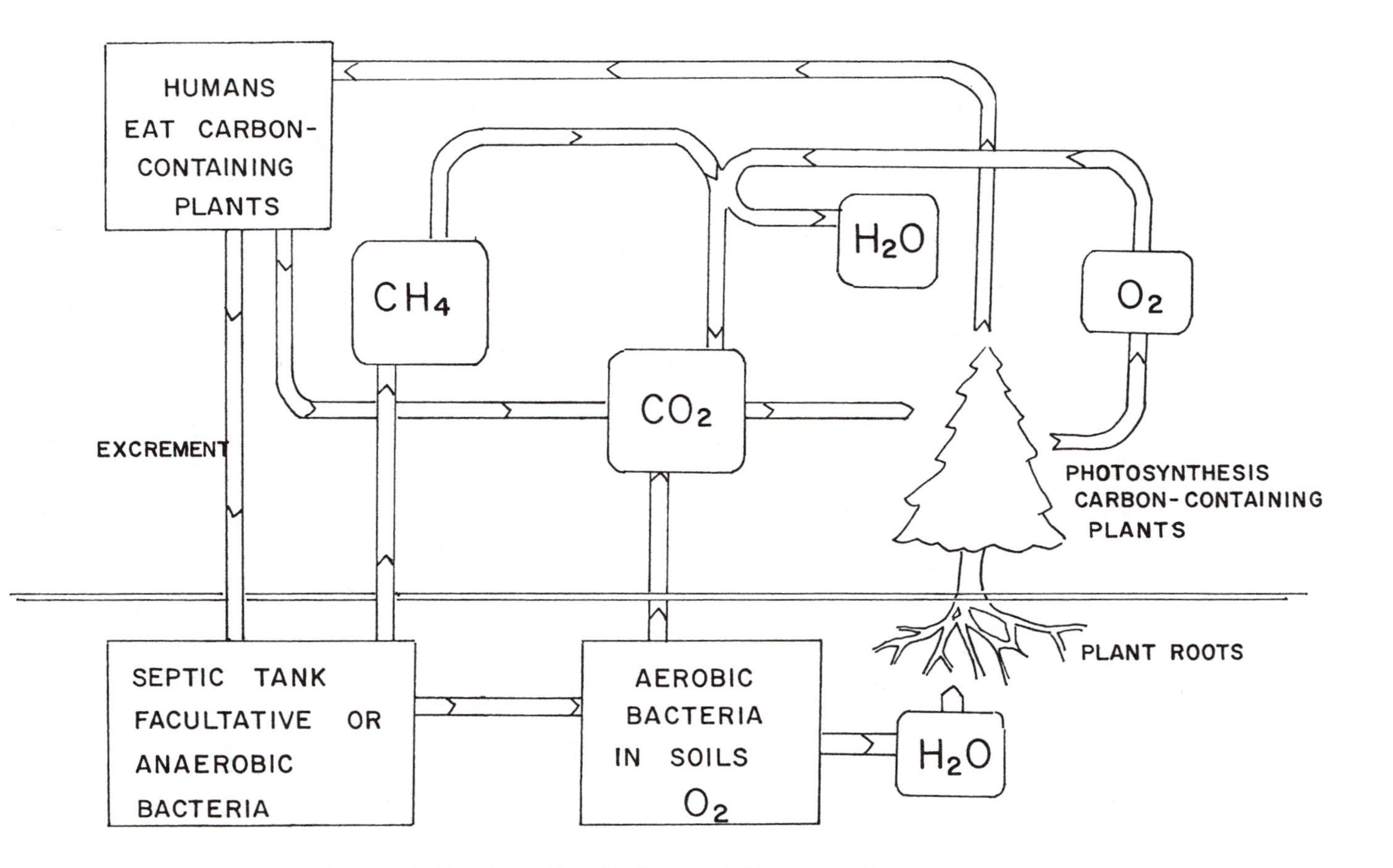

Figure 3.11 - Recycling Carbon and Oxygen in the Ecosystem.

substances, some under aerobic conditions and others under anaerobic conditions. Fungi decompose plant tissues and are generally found near the surface, where aerobic conditions exist.

The extent and type of microbial growth in soil depends upon the same factors that control growth in aquatic environments including:

- nutrient concentrations
- moisture regime
- oxygen availability
- temperature and pH.

Bacteria are the dominant group of microorganisms in the soil, but because their cells are small—many of these organisms are smaller than 0.3 μm—the total biomass of bacteria is frequently less than the mass of fungi. In terms of metabolic activity, certain processes in well-aerated soils are performed by bacteria, and anaerobic bacteria are chiefly responsible for biochemical changes further underground where less oxygen gas is available. The bacterial community in the soil is dominated by twenty-five types; major species are members of *Arthrobacter, Pseudomonas, Agrobacterium, Alcaligenes, Glavobacterium, and Bacillus.*

Bacteria can carry out many processes. Heterotrophic bacteria can metabolize and mineralize many low-molecular-weight organic compounds under a wide range of conditions. They can even decompose some synthetic organic molecules. They can oxidize and reduce forms of nitrogen, sulfur, iron, and manganese. Oxygen and carbon cycle through the ecosystem often with the assistance of microorganisms, as shown in Figure 3.11.

Fungi are especially important in soil because they help to decompose resistant compounds, especially those not readily utilized by bacteria. The filamentous fungi decompose organic nitrogen compounds that are abundant in soil and convert them to ammonium. Fungi are also important in the decomposition of cellulose compounds that are abundant in higher plants. Fungi are the dominant organisms in aerated conditions in the breakdown of lignin. They are also significant in forming *humus*, the native organic fraction of soils. The filamentous organisms can bind soil particles together to improve soil aeration and water movement. This promotes root growth.

Algae and cyanobacteria bring about photosynthesis where water is abundant in the soil. They usually appear on the surface of the soil. They add organic matter to the soil, release molecular oxygen to the atmosphere and soil, and are the primary colonizers in disturbed environments.

Three types of protozoa dominate in the soil: amoeba, flagellates and ciliates. All are mobile, and their concentrations vary appreciably from place to place.

Protozoa are predatory and usually eat bacteria; sometimes protozoa coexist with bacteria, depending upon the conditions. Protozoa serve an ecological service: they limit and maintain healthy bacterial populations.

NATURAL WATER MICROBIOLOGY

Many forms of microbial life can exist in water provided the correct physical and nutritional requirements for growth are met. Dissolved oxygen is necessary for the growth of aerobic bacteria and protozoa. Nitrogen, phosphorus, and light are essential to the growth of algae. The numbers and types of microorganisms present indicate water quality. In "clean" water with a low nutrient level, the total number of microorganisms is limited, but a great variety of species can exist. As the nutrient content increases, the total number of microorganisms might increase, but the variety of species is reduced. In a polluted, anaerobic stream, a few species of anaerobic or facultative bacteria will predominate. Typical numbers of bacteria for various waters are presented in Table 3.1

Table 3.1
Typical Bacterial Counts in Waters

Source	Bacteria per 100 ml	Coliform Bacteria per 100 ml[a]
Tap Water	10	0 - 1
Clean, natural water	10^3	0 - 10^2
Polluted Water	10^6 - 10^8	10^3 - 10^5
Raw Sewage	10^8	10^5

[a] Coliform bacteria are present in sewage but die off with time in natural waters. Their natural habitats are the intestines of warm-blooded mammals and the soil. *E. Coli* populations have a half-life of 15 hours.

INDICATOR ORGANISMS

Water used for drinking or bathing can serve as a vehicle for the transmission of a variety of human *enteric pathogens* that cause waterborne diseases. Enteric pathogens refer to those bacteria, viruses, and protozoa that are found in the intestinal tract and excreted in feces. The detection of pathogens in water is difficult, uneconomical, and impractical in routine water analyses. Instead, water is tested using a surrogate indicator of fecal contamination. Because nonpathogenic microorganisms also inhabit the intestines in large numbers and are always present in feces with any pathogens, these surrogates may be used as indicators of fecal contamination.

The main characteristics of a good indicator organism are:

- its absence implies an absence of any enteric pathogens;
- its density is related to the probability of the presence of the pathogens; and
- its survivability in the natural environment is longer than the pathogens.

No ideal indicator organism exists, but certain bacteria, such as *coliform, streptococci*, and *Clostridium perfringens*, are regarded as evidence of fecal contamination and have been used for assaying water quality for many years.

The total coliform group of bacteria is the one most commonly used. The coliform group comprises several genera of bacteria belonging the family *Enterobacteriaceae*. Among the many microorganisms included in this group are *Escherichia coli, Enterobacter, Klebsiella, Serratia* and other related bacteria. These bacteria include, by definition, "all aerobic and facultative anaerobic, gram-negative, non-spore-forming, rod-shaped bacteria that ferment lactose with gas formation within 48 hours at 35°C." [9]

In 1884, Danish scientist Hans Christian Gram (1853-1935) developed a method of staining bacteria so that they could be viewed under a microscope; the method is called the Gram stain. Bacteria are classified into two types, depending upon how they react to the application of the stain and subsequent alcohol treatment. Some types of bacteria that are treated with the stain (crystal violet) and then alcohol lose the dye's color; those types are gram negative. Bacteria that retain color after the alcohol treatment are gram positive.

In drinking water, where no coliform should be present, the total coliform test is used as an indicator of recent fecal pollution. In polluted streams, sewer outfalls, and swimming areas, *fecal coliform* are counted by using the elevated temperature test (44.5°C $\pm$0.5). *E. coli* is the most frequent and predominant type of coliform found in the human intestine, but has a half-life of only 15 hours after leaving a host's body[10]. Other members of the coliform group, usually found in soil and on vegetation, may also be encountered in feces, but in low numbers. In tropical countries, *E. coli* is not the predominant intestinal coliform, so the total coliform test, rather than the fecal coliform test, is a more useful measure of pollution.

Fecal streptococci, another type of intestinal bacteria — commonly used as indicator organisms in European countries — are more common to animals than to humans. These bacteria are frequently counted in conjunction with fecal coliform. The ratio of fecal coliform to fecal streptococci (FC/FS) is used to differentiate the sources of pollution. With a ratio of 4.0 or more, the pollution is considered from human wastes, whereas ratios of less than 0.7 indicate pollution from animal wastes.

Methods To Determine Bacterial Concentration

Bacterial contamination is quantified by three test methods, each of which is described below: the Multiple-Tube Fermentation Test (MTFT) (also called the Most Probable Number (MPN) test the Membrane Filter Test (MFT), and the Heterotrophic Plate Count Test (HPC).

Multiple-Tube Fermentation Test (MTFT). To perform this test, the water under study is diluted "serially." Serial dilution involves taking 1 ml of sample and diluting it to 10 ml; then 1 ml of the resulting solution is diluted to 10 ml. This process continues until there are four diluted samples of these strengths: 0.10, 0.01, 0.001 and 0.0001. After the dilutions are completed, five 1 ml samples of each dilution are added to *fermentation tubes*, test tubes that contain a culture medium (a substance that will support the growth of bacteria). Attached to these tubes are gas-collection tubes. The samples are heated to 35 °C for 24 hours in a water bath. If gas is found in the gas-collection tube, bacteria are presumed to exist in the sample. For each dilution, the number of positive (gas-collected) results are reported. Because there were five tubes for each dilution, the number is reported as a fraction of the five, so two positive results would be reported as 2/5. Figure 3.12 shows a technician examining fermentation tubes for signs of gas formation.

Figure 3.12 - Examining Fermentation tubes for signs of gas generation.

The most probable number (MPN) of bacteria is calculated from the results of the MTFT. This number, the MPN, is a statistical estimate of the number of bacteria present, not a quantitative determination. The MPN/100 ml of sample can be determined by the *Thomas Equation*:

$$\frac{MPN}{100\ ml} = \frac{Number\ of\ positive\ tubes\ X\ 100}{\sqrt{(mL\ of\ sample\ for\ negative\ tubes)\ X\ (mL\ of\ sample\ for\ all\ tubes)}} \tag{3.7}$$

Membrane-Filter Test (MFT). The Membrane-Filter Test (MFT) is highly reproducible, yielding numerical results faster than the MTFT procedure. The sample under study is filtered through a membrane that has holes in it smaller than bacteria. The filter is then put onto an *agar*-filled petri dish and incubated from 20 to 24 hours, depending upon the specific test procedure. (Agar is a good nutrient medium for the growth of bacteria.) The number of bacterial colonies that are found are counted and reported against the amount of water sample filtered. Figure 3.13 shows a Petri dish with bacterial colonies growing in the agar medium.

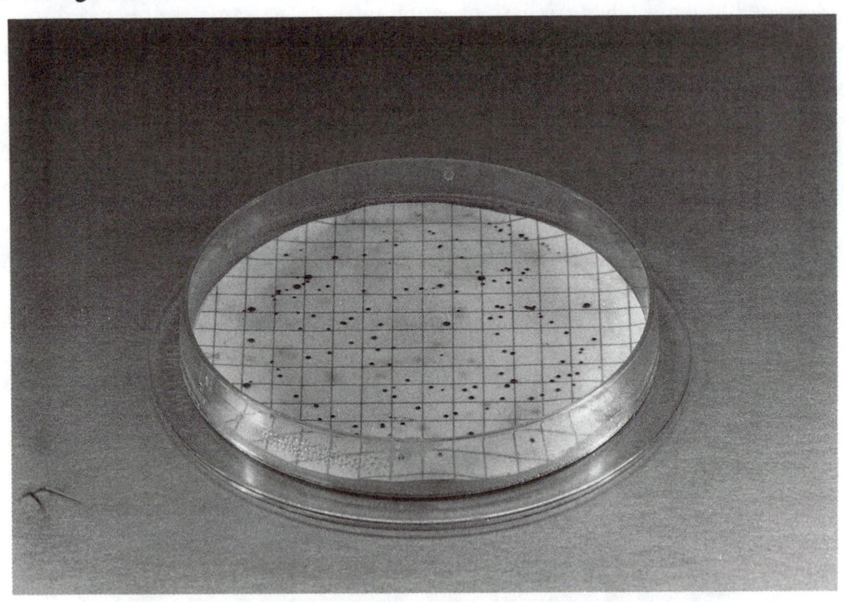

Figure 3.13 - Petri dish with bacterial colonies growing in it.

Heterotrophic Plate Count (HPC). This test, previously called the Standard Plate Count (SPC), is a way to estimate the number of live heterotrophic bacteria in water. There are three different methods outlined in *Standard Methods*[11]: Pour Plate Method, Spread Plate Method, and Membrane Filter Method. Each has advantages and disadvantages, given the condition of the water. In all cases, water samples are added to a plate of agar and incubated The number of colonies arising from the incubation are counted. Each of these tests is simple to perform and can be conducted by minimally-trained personnel.

Plus, the results of these tests are available quickly, an advantage when assessing a swimming pool or beach.

Sterilization and Disinfection

Just as conditions can be created for the optimum growth of bacteria, so unfavorable conditions can be used to eliminate bacteria. Complete destruction of microbial life is called *sterilization*. Complete sterilization is usually achieved by the application of moist heat, which may not always deactivate bacterial spores. *Disinfection*, on the other hand, implies the destruction of pathogens (disease-causing bacterial) only. There are physical and chemical methods to achieve disinfection. Some physical methods are moist heat, dry heat, cold, filtration, and radiation. Chemical agents that achieve disinfection are phenols, alcohols, halogens (such as chlorine), soaps, detergents, alkylating agents (like formaldehyde and glutaraldehyde) and some heavy metals (mercury in merthiolate and silver in silver nitrate).

References

1. Cano, R. Colome, J.S. *Microbiology.* West Publishing Company. San Francisco, 1986. p. 351.

2. Mudrack, Klaus and Kunst, S. *Biology of Sewage Treatment and Water Pollution Control.* John Wiley & Sons, N.Y. 1986, p. 99.

3. Hammer, Mark J. *Water and Wastewater Technology.* John Wiley & Sons, Inc. N.Y., 1977, pp. 59-61.

4. Mudrack, p. 95.

5. Gerardi, M.H. *et al.*, "An Operator's Guide to Wastewater Viruses." *Public Works*, 50. 1988.
5. Mudrack, p. 129

6. Mudrack, p. 58.

7. Metcalf & Eddy. *Wastewater Engineering Treatment, Disposal and Reuse*, Third Edition. McGraw-Hill Publishing Company, N.Y., 1991, p. 369.

8. Bolton, R.L. , Klein, L. *Sewage Treatment — Basic Principles and Trends.* Ann Arbor Science Publishers, Inc. Ann Arbor, MI, p. 100.

9. Greenberg, A.E., Clesceri, L.S., Eaton, A.D. *Standard Methods for the Examination of Water and Wastewater*, 18th Edition. American Public Health Association, Washington, D.C. 20005, 1992. p. 9-45.

11. Water Pollution Control Federation, *Wastewater Biology: The Microlife.* Water Environment Federation, Washington, D.C., 1990, p. 123.

10. Greenberg, p. 9-32.

CHAPTER 4

SOIL AND SITE EVALUATIONS

Topics of This chapter:

- *Soil Science Principles*
- *Factors Influencing the Siting of an Onsite Wastewater Treatment System*
- *Steps to Conduct a Soil and Site Evaluation*

REVIEW OF SOIL SCIENCE PRINCIPLES

Soil science comprises principles and conventions for conducting site evaluations. By using these principles and conventions, designers and installers can use a common language to describe soils. This chapter will describe the methods and conventions soil scientists have developed for performing soil evaluations.

Unless otherwise noted, these methods and conventions were developed by the United States Department of Agriculture, Soil Conservation Service, and detailed in the *Soil Survey Manual*, which the agency produces. This text is based on Chapter 4 of the 1984 edition.

THE ORIGINS OF SOILS

Soil is the part of the Earth's surface that is made up of disintegrated rock and humus. Soils on a particular site can be fairly uniform or can show a dramatic difference in character over a short area. The coastal area of Florida has soils consisting mostly of sand and a lot of Texas soils are clayey. But Florida and Texas are unusual because most United States areas have soils that are composed of several different layers called *horizons*. The thickness and composition of these horizons are what distinguishes one soil profile from another.

Soil Formation

Soil develops through the interaction of climate, topography, time, parent material, and presence of biological activity. Each contributes to the character of a soil in its own way. The total effect of these factors is the disintegration of minerals into smaller and smaller pieces through the general process known as *weathering*. Weathering can result from three processes: mechanical disintegration, chemical processes, or biological activity. Mechanical disintegration is the process of breaking rock mass into smaller particles. It can result from climatic effects or through movement, the rocks rubbing against each other or by water or wind moving the rocks[1].

Chemical reactions include: reactions of the original materials with water, oxygen, or organic material; dissolution in water; and secondary reactions with other chemicals. Chemical reaction of the original materials is the beginning of the process, which leads to new products and reactions. In the presence of oxygen, some materials oxidize to new products. Some of these products may be soluble in water and become transported elsewhere. Water can bind directly with minerals in a process known as *hydration*. Other minerals react with the acids produced by living organisms, each reaction leading to the formation of new products. Sequential reactions of products and byproducts create the character that is distinct for each soil.

Many times, lichens attach themselves to the surface of rocks and, as part of their metabolism, secrete acid. This acid begins to break down large rocks and weaken the structure of the rock over time. Once a fissure opens in the rock, water can enter the crack, undergo cycles of freezing and thawing, and break the rock into fragments[2].

Soil Porosity

Void spaces in soil are called *pores*, and the total volume of these voids is the *porosity* of the soil. The character of the porosity is different among soil types and is important to consider when defining the soil for use in an onsite system. The porosity of a soil is expressed as a percentage of pores to the total volume of the soil, without considering the presence of any material, such as water or gases, that may occupy those pores from time to time. The presence of both air and water in the pores, as well as many other factors, defines what kinds of microbial activity can take place in the soil[3].

Soil Air

Air in the pores is usually different in composition from the air above the soil due to the many biological and chemical processes that occur in the soil. Also, the soil humidity approaches 100 percent; the carbon dioxide content is often several hundreds of times higher, and the oxygen content lower than atmospheric air[4]. Soil air occupies the larger pores and water fills the smaller pores. The degree of purification that onsite systems can achieve is highly dependent upon the amount of oxygen available in the soil to allow biological process to oxidize the organic material in the effluent. Smaller-grained soils usually are poorly aerated, as is the case with clay, and cannot replenish oxygen as fast as it is consumed.

Soil Particle Surface

Chemical reactions can also occur on the surface of the soil particles themselves, particularly the finer textured soils, such as clays. Many clay particles are plate-like and have strong negative charges along their broken surfaces. These charges can both attract and repel other substances, depending on the charge on the other substances. The adhesion of the other substances, which can include chemicals or microorganisms is called *adsorption*. Once adsorbed on the clay particle, the chemical or microorganism can interact with other substances as they flow by. For example, viruses become food for passing predators or can be inactivated by exposure to hostile environments. Some chemicals may participate in additional reactions. Ammonia, for example, has a strong positive charge and quickly adsorbs to clays. However, ammonia can quickly become oxidized to nitrate, which is negatively charged and will be released by the negatively-charged soil particle[5].

The Soil Profile

In general, soil is composed of several layers, which are called *horizons*. The sequence of these horizons is called the *soil profile*. Each horizon has a distinct character, and it is important to characterize each layer if an onsite system is to be properly sited and installed[6].

Most soil will have at least three horizons: the topsoil, the subsoil, and the parent material. Some soils have only two horizons, a topsoil and parent material. Figure 4.1 shows the profile of a typical soil having three horizons.

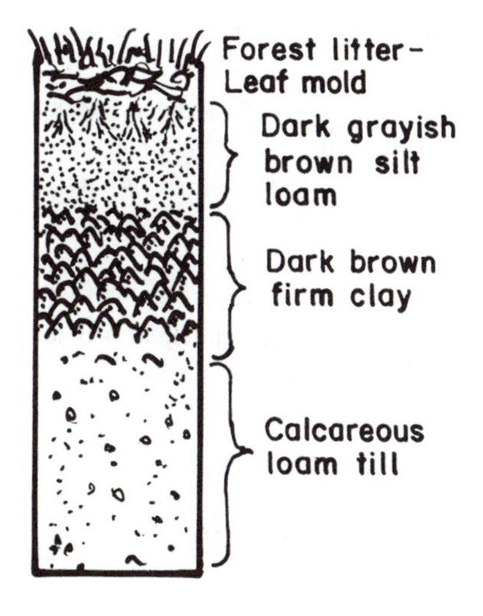

Figure 4.1 - A Soil Profile

The A Horizon. The surface horizon is called the A horizon and is commonly called *topsoil*. Technically, the A horizon is composed of soil mineral mixed with decomposed organic material, called *humus*. The A Horizon is a zone of accumulation; that is, the A Horizon results from the deposition of material, whether it is organic or inorganic. The thickness of this layer ranges from 50 mm to 600 mm (2 in to 2 ft).

The B Horizon. The B Horizon is next and is called the the *subsoil*. This horizon is made up of one or more layers of soils composed of the minerals from the parent material below it and the material deposited above it. B horizons tend to be dominated by certain minerals and elements. Common elements include iron, calcium, manganese, and aluminum, and silicon, including quartz, carbonates and clays. Since this horizon is chemically different from the surrounding layers, full characterization is essential. The B Horizon generally has little organic material. Some B Horizons, however, can contain much organic material.

The C Horizon. The C horizon is distinguished from the others by the a lack of weathering. In concept, the C horizon is what the higher level horizons were before the process of weatherization.

Besides the three major horizons, some soils exhibit one or two other horizons: the O horizon and the E horizon.

The O Horizon. The O horizon is an organic layer, consisting of leaves, plant remains, and animal droppings that sometimes lies above the A horizon. The O horizon would be found in a grassland, woods, wetland, bog, or other location where organic material accumulates.

The E Horizon. The E horizon is a horizon that develops directly below the A horizon and results as minerals and elements are leached out of the A horizon. Leaching occurs as water percolates through the horizon, dissolving the minerals in it, so this layer is sometimes called the *zone of leaching*.

The major horizons can be further divided by assigning suffixes to them. Such suffixes include designations for wetness, climatic environment, vegetation and whether the soil has been disturbed by forces such as plowing. For example, a horizon designated as *Aca* indicates some calcium carbonate or chalk accumulations, and *Bn* indicates humus accumulations in the B layer.

Hardpan describes a buried, hard, impervious layer in the A, B or C horizons. Shallow hardpans are called *duripans*. *Fragipans* usually start 400 to 600 mm (1.5 to 2 ft) below the soil surface and fade away at 1.2 to 1.5 m (4 to 5 ft) below the surface and occur mainly in loamy soils that are leached of their carbonates; fragipans are slowly permeable and contribute to a locally high, or *perched*, water table. Perched water tables pose significant challenges for siting an onsite system.

Classifying Soils

Soil is a mixture of materials that can be distinguished by the size. The materials can be divided in the rock fragments and fine earth materials as shown in Table 4.1. Rock fragments are particles equal to or greater than 2 mm (1/16 in) in diameter. The fine earth fraction consists of all particles less than 2 mm in size.

Rock fragments can be further distinguished by sizes: *boulders*, *cobbles*, and *gravel*. A rule of thumb that is used, not necessarily by soil scientists, is that if the percentage of rock fragments exceeds 50 percent of the total soil mass, the soil is considered *bedrock*. If the rock fragments comprise less than 50 percent of the mass, the soil is considered "cobblely" or "gravelly," depending on the type of rock fragment dominating the soil. The fine earth fraction is the concern of the soil evaluation if the soil is not considered bedrock. Like the rock fragments, the fine earth fraction is divided into categories, depending upon the size of the particles: sand, silt, and clay.

Table 4.1
Soil Size Classification

Rock Fragments	
Class	**Size**
Boulder	> 300 mm
Cobble	150 - 300 mm
Gravel	2.0 - 150 mm
Fine Earth Fraction	
Class	**Size**
Sand	0.05 - 2.0 mm
Silt	0.002 - 0.05 mm
Clay	<0.002 mm

Table 4.2
Sand Classification

Sand Subset	Size Range (mm)
Very Coarse	1.00 - 2.00
Coarse	0.50 - 1.00
Medium	0.25 - 0.50
Fine	0.10 - 0.25
Very Fine	0.05 - 0.10

Sand is loose and composed of single grains of minerals. Particle sizes for sands range from 0.05 to 2.0 mm. Sand particles are predominantly quartz (SiO_2), and other minerals including feldspar, hornblende and mica. Sand is further divided into *very coarse*, *coarse*, *medium*, *fine*, and *very fine* as shown in Table 4.2

Silt and clay are the smallest soil particles, and their combination is what provides the greatest treatment variation in soil. The combination of sand, silt, and clay in a soil is called *loam*. Common soil textural classes are:

- Silt is composed of single particles that tend to hold together, especially when moist. Silt Particles range from 0.002 mm to 0.05 mm.
- Clay is composed of particles invisible to the naked eye and which stick together whether wet or dry. Clay particles are under 0.002 mm.
- Sandy loam is one that is mainly sand but has enough silt and clay to make the soil stay together. Sand grains can be distinguished, and the soil can be molded.

- Silt loam has at least 50 percent silt. Silt loams appear as clods and appears "floury" when dry.

- Clay loams are at least 30 percent clay and exhibit "plastic" characteristics.

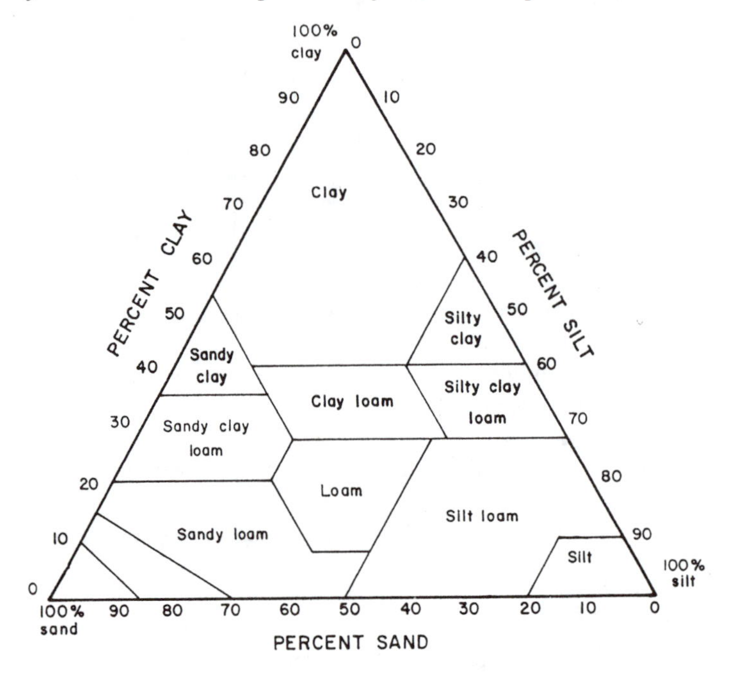

Figure 4.2 - USDA Soil Triangle

Figure 4.2 illustrates the relationship among the *separates* of soil particles and the classifications of soils which combine different portions of each type. This is the USDA Soil Triangle and shows the interrelationships among sand, silt and clay.

Soil Color

Texture alone is insufficient to fully characterize a soil. Soil color is also integral to soil characterization and, within limits, can be used to identify important soil properties. Darker colors may show the presence of organic matter. Gray colors may indicate an absence of oxygen either from water saturation or other factors. Mottles, alternate streaks of oxidized and reduced materials, suggest periods of oxidation and reduction that indicate seasonal saturation or imperfect drainage. Evidence of soil saturation is very important in considering the type of onsite wastewater system for a particular site.

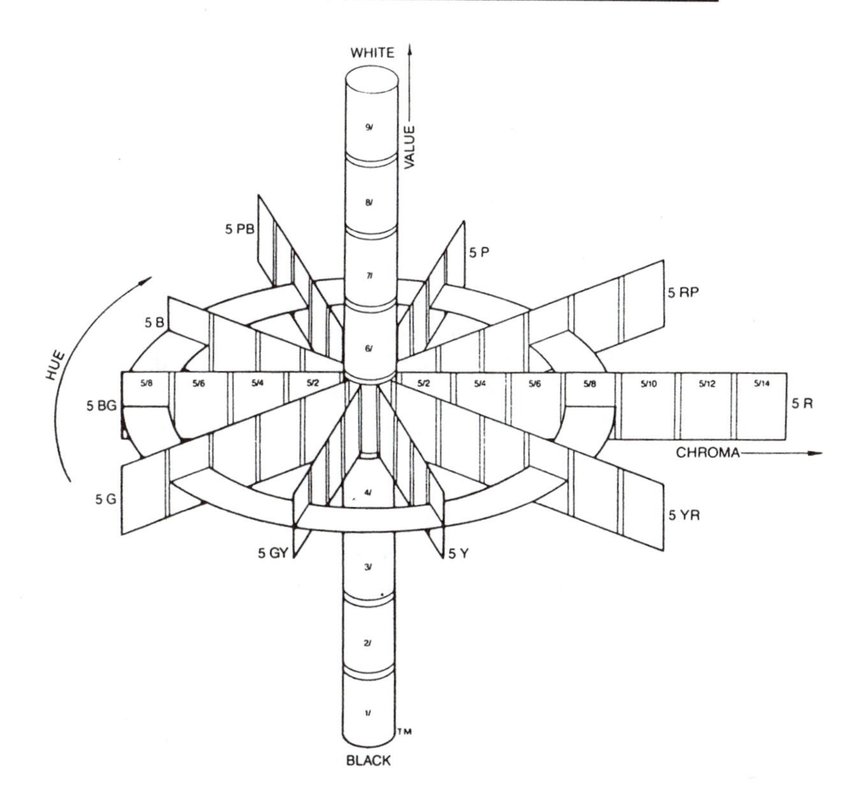

Figure 4.3 - The Munsell Color System

Soil color is established by comparing a soil sample to a standard color chart **in natural light**. The most common color system is the Munsell Color System, which consists of 175 standard colors that encompass most soil conditions.

Munsell Color System[a]. The Munsell color system uses three elements to color to distinguish a soil: hue, value, and chroma. Figure 4.3 shows how the Munsell Color System classifies colors according to these conventions.

Hue is spectral quality of the color in comparison to five principal colors: red, yellow, green, blue, and purple. There are another five intermediate hues that identify the midpoints between the principal color pairs. Hues are divided into four equal gradations between 2.5 and 10. These gradations are prefixes used before the color or color-pairs.

---

[a]The Munsell Color System is produced by Munsell Color, a division of Macbeth Corporation, New Windsor, NY

Value is the lightness or darkness of the color in relation to gray on a scale from 0 to 10. Black has a value of 0; white has a value of 10. Medium gray has a value of 5.

Chroma is the intensity of a color on a scale from 0 to 14. Colors diluted by grey have a low value; pure colors have a high chroma.

Soil colors can be judged accurately on sunny days when the sun is high. Clouds and the horizonal sunlight distort the light, making comparisons difficult. Clouds make colors appear more neutral while the sunlight penetrating low over the horizon is redder than when shining from directly above. Like soil texturing, judging colors is a learned skill. Experienced evaluators can judge a soil within one gradation of hue and one gradation of value and chroma.

Soil Color Patterns. Soils typically exhibit several colors that form a unique pattern. The nature and color of this pattern can indicate the minerals in the soil, degree of wetness of the soil, and the amount of oxygen in the soil. Usually there is a dominant color to the soil. Intermixed with this may be other colors. The dominant soil color should be noted and other colors identified as such.

Mottles are spots or streaks of differing color that appear in the soil. Mottles resulting from periods of saturation are the most important for siting an onsite system. These mottles may be either red, yellow, orange, or gray, depending on whether iron was exposed to or deprived of oxygen. Soils deprived of oxygen and subject to leaching have gray mottles. While mottles are usually described as spots, they may appear as streaks or other shapes, which should be noted.

Mottles are described in terms of four qualities:

• *Quantity* • *Size* • *Contrast* • *Color*

Quantity is a general estimation of the amount of the surface the mottles cover.

Few: Less than 2%

Common: Between 2% and 20%

Many: Greater than 20%

Size is the approximate diameter or longest dimension.

Fine: Less than 5 mm

Medium: From 5 mm to 15 mm

Coarse: Greater than 15 mm

Contrast refers to the difference among the soil colors.

Faint mottles are difficult to see because they differ from surrounding colors by no more than 2.5 units of hue, two units of value and one unit of chroma in the Munsell Color System.

Distinct mottles are better seen because of greater contrast. They usually differ from surrounding colors in one of two ways. Some mottles have the same hue but value may vary three or four units, and the chroma may vary two to four units. Conversely, some mottles have hues that vary by 2.5 units but have value and chroma that vary by at least two units of value and one unit of chroma.

Prominent mottles are marked by contrasting colors. There are specific formulas for classifying them, but a good rule of thumb is that mottles are considered prominent if they are an outstanding feature of the horizon.

Gleyed refers to soils whose development was affected by strongly reducing conditions. These conditions result in soils with a bluish-gray color that results from iron compounds being "reduced," that is, deprived of oxygen. Gleyed conditions are important because the lack of oxygen in the soil can indicate saturated conditions - a typical limiting factor in an onsite system siting. One sign of gleyed conditions is a color change when the soil is exposed to air: gleyed soils having a high iron content will turn brown when exposed to oxygen.

Shape is an important consideration in describing mottles because the shape of mottles is related to the structure of the soil that formed the mottles. Shapes may be identified as how they appear to the observer: "spots," "bands," or "streaks" are common descriptions.

Redoximorphic Features. The U.S. Soil Conservation Service has recently adopted more exact terminology for describing the color changes that are produced in soils that are saturated part of the time when little oxygen is available. The colors that result from the reducing (non-oxygenated) environment are called redoximorphic features and indicate that the soil does not provide the aerobic conditions that are desired for adequate effluent treatment[7]. The term *gleyed* most closely describes a redoximorphic feature.

Soil Structure

Like any other material, soil is composed of basic structural elements. For a sand type, the basic elements are the sand grains, thus the description *single grained*. Under certain conditions finer-textured soils can be viewed as a single mass, thus its description is *massive*. Both sands and massive soils can be considered *structureless*. Other soils form structural elements called *peds*, which are the natural physical organization of a soil or the lumps that soils generally break into. Peds can be distinguished because they will break along surfaces that are naturally weak and in a recurring pattern that can be seen across soil samples. Each ped is an aggregate of thousands or millions of soil particles.

Ped qualities have a marked effect upon the suitability of a soil for a system. Consequently, peds must be understood in detail with respect to three significant properties: grade, size and shape.

Grade refers to the ease with which peds can be separated.

- *Weak* describes peds that are not easily observed and may be confused with structureless soils. When separated, the peds may fall apart or not exhibit planes of weakness.
- *Moderate* describes peds that are well formed but may break apart when separated; planes of weakness are evident.
- *Strong* describes peds that are easily identified and separated from each other.

Size refers to the size of the peds, according to the convention *very fine*, *fine*, *medium*, *coarse*, and *very coarse*. The convention is applied differently to peds, depending also on the shape of the ped. The conventions are shown in Table 4.3.

Table 4.3 - SCS Size Convention for Peds

Size of ped	Symbol	Granular, Crumbly, Platy	Angular, Subangular Blocky	Prismatic Columnar
Very Fine	vf	< 1 mm	< 5 mm	< 10 mm
Fine	f	1-2 mm	5-10 mm	10-20 mm
Medium	m	2-5 mm	10-20 mm	20-50 mm
Coarse	c	5-10 mm	20-50 mm	50-100 mm
Very Coarse	vc	> 10 mm	>50 mm	> 100 mm

Shape describes how the peds are formed. The basic ped shapes are:

- *Platy* peds are just that - flat, like plates, and generally horizontal and overlapping.
- *Prismatic* peds tend to be long and have flat vertical faces. Their tops are generally flat.

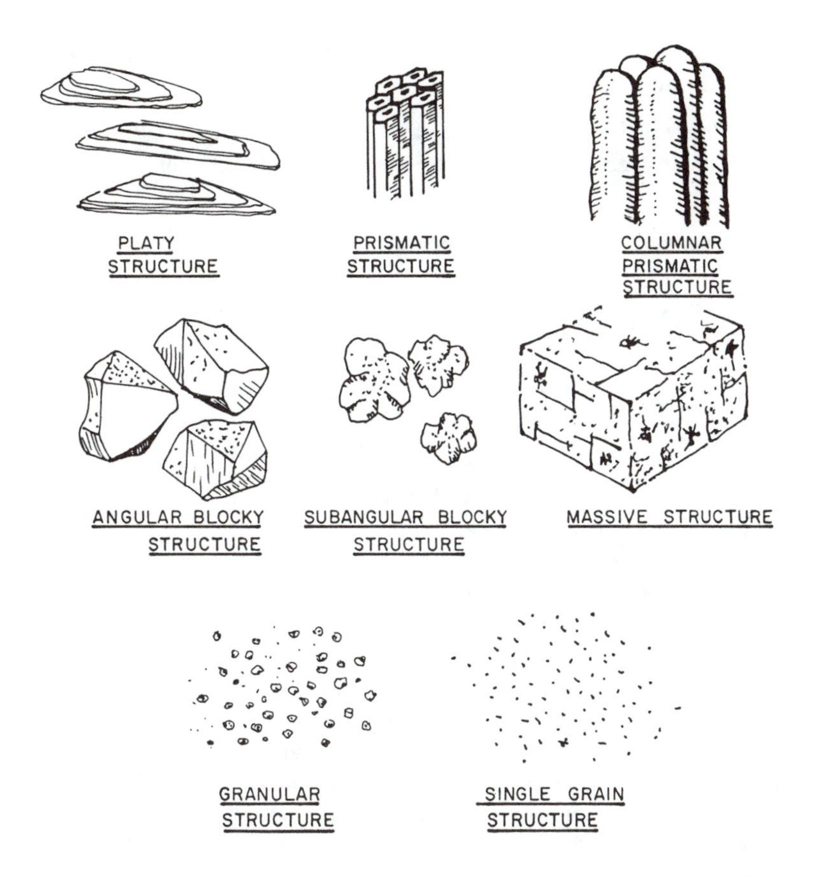

Figure 4.4 - Shapes of Peds

Figure 4.5 - Health Inspector from New Canaan, CT, examines peds

- *Columnar* peds are similar to prismatic peds but tend to have distinct, rounded tops.
- *Blocky* peds are just that: blocky. Blocky peds can be **subangular blocky** if their faces are rounded or **angular blocky** if their faces have sharp edges.
- *Granular* peds have the appearance of grains.

Peds must be described as they appear in their natural state. Figure 4.4 shows examples of each of the ped shapes that has been described.

Soil Moisture

Soils look and behave differently, depending on the amount of water in the pores between the soil particles. The amount of water the soil can hold is related to the makeup of that soil. It is important to be able to estimate the moisture in the soil to both identify the soil and to estimating its ability to transmit water. Soil can retain moisture because of attractions between the soil particles and water molecules. This attraction increases as soil particle size decrease. Sands retain little moisture because the individual pores are large, and this size limits the contact area between the water and the sand grains. Clays, on the other hand, are composed of tiny particles and, consequently, small pores. As a result, clays have a large ability to retain water.

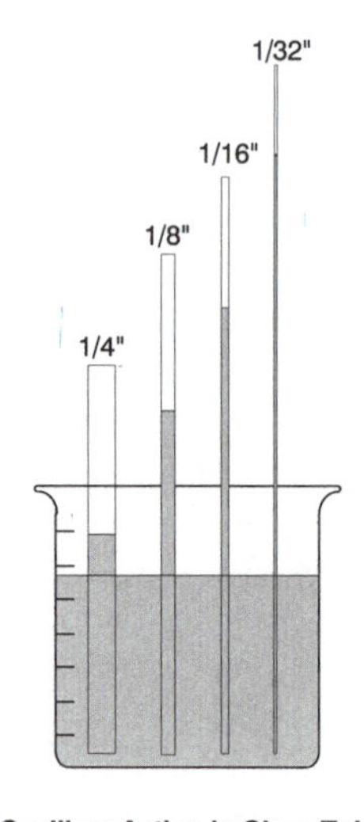

Capillary Action in Glass Tubes

Figure 4.6
Capillary Action

This effect, *capillary action,* can be seen by placing glass tubes of different diameters in a glass in water, as shown in Figure 4.6. As the tube diameters get smaller, the height of the water column increases. This happens because, in the smaller tubes, there is a larger surface area of the glass compared to the surface area of water. This capillary action also occurs in soils, so smaller-sized soils hold more water. This phenomenon is described more fully on page 91.

Soil moisture is measured in the amount of *pressure* the attractive forces in the soil particles exert on the water. As the pore spaces fill with water, the attractive forces diminish. When the soil is saturated, no pressure exists, and water will move along a gradient, depending on the other characteristics of the soil. The unit of soil pressure is the *Kilopascal* (KPa, 1,000 Pascals (Pa)) and in pounds per square inch (psi) in U.S. customary units.

Soil moisture can be estimated in the field using one of three general conditions: dry, moist, or wet. *Dry* soils exert a negative pressure (tension) of at least 1,500 Kpa (218 psi). They feel dry to the touch. Finer textured soils that are dry may be in hard clumps. *Moist* soils have a soil tension of between 1 KPa and 1,500 KPa (0.15 and 118 psi)[8]. They feel moist to the hand but will not release water if squeezed. *Wet* soils, on the other hand, will exhibit free water when squeezed[9].

Water moves through soil by either tension or gravity. The movement of this water is affected by the characteristics of the soil, which may vary greatly. *Hydraulic conductivity* or *permeability* is the ability of water to move through a soil. When soil is *unsaturated*, that is, does not have pore spaces filled with water, water is pulled through the soil mainly by the attractive forces of the soil particles. Water moves through saturated soil simply because gravity pulls the water along. As the saturated zone becomes deeper, the water exerts pressure on the soil, increasing the movement of the water until the soil reaches its *saturated hydraulic conductivity*, which is the ability of water to pass through saturated soil[10].

Table 4.4
Saturated Hydraulic Conductivity Classes

Conductivity Class	Conductivity (min/mm)
High	
Very High	<0 .166
High	0.166 - 1.66
Moderate	
Moderate	1.67 - 16.65
Moderately low	16.65 - 166.53
Low	
Low	166.53 - 1665.30
Very Low	> 1665.30

Hydraulic conductivity is estimated by the time period a soil remains wet. Highly conductive soils will stay wet for just a few hours. These soils are characterized by large particles and large, connected pore spaces. Highly conductive soils have little silt or clay. Moderately conductive soils may remain wet for several days. These soils are characterized by larger percentages of silt and clay. These soils typically form cracks when dried. In some soils these cracks may remain even if the soil is rewetted. Slowly conductive soils will remain wet for a week or more after wetting. They usually have massive, platy structure, or few connected pores. Hydraulic conductivity is not directly related to porosity because sometimes clays have higher porosity than sands, but are much less permeable. The classes of hydraulic conductivity are listed in Table 4.4[11].

Consistence. Consistence refers to the combination of soil properties that determine its resistance to crushing and its ability to be molded or changed in shape. Such terms as *loose*, *friable*, *firm*, *soft*, *plastic*, and *sticky* describe soil consistence. Consistence also refers to how well the soil particles stick to each other and to other substances. Soil consistence will vary with soil moisture, so consistence is evaluated differently, depending on whether the soil is moist or dry.

Boundaries. Soil horizons vary in how dramatically they change. Some horizons may be distinct from others, separated by an easily-visible change in soil characteristic; others show gradual change. Understanding and describing how the boundaries between soil horizons change provides more informationup upon which to base a decision for the siting and design of an onsite system. Boundaries are measured in terms of the transition width between adjacent horizons. These widths are shown in Table 4.5.

Table 4.5
Boundary Transition Widths

Class	Symbol	Width
Abrupt	(a)	< 25 mm
Clear	(c)	25-65 mm
Gradual	(g)	65-130 mm
Diffuse	(d)	> 130 mm

Roots. The presence or absence of plant roots is important because the roots can affect soil colors and hydraulic conductivity. Roots can also be used to indicate soil condition because roots cannot penetrate firmly packed soils and cannot grow under water-saturated, low oxygen conditions.

Roots can be described by simple classes of Quantity and Size. A soil has *few* roots if the roots take up less than one percent of the surface area. A soil has *common* roots if roots take up between one and five percent of the surface, and a soil has *many* roots if roots take up more than 5 percent of the area. Roots are considered *very fine* if they are less than 1 mm in diameter, *fine* if they are between 1 mm and 2 mm in diameter, *medium* if they are between 2 mm and 5 mm, and *coarse* if they are greater than 5 mm in diameter.

Topography. Topography is the land's surface contour and is sometimes called *relief.* The interaction of topography and natural forces at a site will affect the distribution and type of soils. For example, a sloping site may have more top soil near the crest and base than along the slope where soil may have been eroded by wind and precipitation. While possibly similar, soils at the base may be wetter than soils at the crest because saturation often occurs at the bottom where water collects. On sloped areas, soils parallel to an elevation contour will tend to be more similar than soils perpendicular to the contour.

Surface topography can be described according to four conditions:

- Smooth: The boundaries are flat.
- Wavy: The boundaries have channels or alternating contours.
- Irregular: The boundaries have no distinct pattern.
- Broken: The boundaries are disjointed.

Use of Soil Survey Maps

Valuable information about a site can be collected by reviewing soil survey maps that are already prepared. The United States Department of Agriculture (USDA) Soil Conservation Service (SCS) has conducted soil surveys for about 95 percent of the counties in the United States, providing information on soils. Soil survey maps are produced by the SCS upon the completion of field surveys conducted by certified soil scientists. Soil survey maps fall into two categories: Reconnaissance and Detailed. Reconnaissance maps cover wider areas and are based on soils sampled in the area. These maps are drawn to scales that include 1:62,500 (1 in equals 1 mi). Detailed maps, which may be drawn at a scale of 1:200,000 (1 in equals 3.17 mi), are based on intensive review and sampling of the study area[12].

Soil survey maps may provide preliminary information, but have limitations. Some surveys were conducted many years ago and do not reflect current conventions and practice. Published soil surveys may not identify *inclusions*, which are soils of a different type that appear in localized areas of another dominant type of soil. Transitional zones may not be marked well, so sites near classification boundaries may have intermediate properties of the identified soils.

Since the SCS was created with agricultural uses in mind, soil survey maps identify six drainage classes—descriptions of the wetness or dryness of the soil—in terms of crop yield and farming operations. While descriptions in terms of crops may seem to be a hinderance to the application of the information, it is actually a benefit. The type of vegetation on a site can yield valuable information about the soils beneath the surface.

"Excessively drained" soils have very high or high hydraulic conductivity. Seasonal saturation occurs at depths of greater than 1.5 m (5 ft).

"Somewhat excessively drained" soils have a high hydraulic conductivity. Seasonal saturation occurs at depths of greater than 1.5 m (5 ft). Somewhat excessively drained soils can support crops; excessively drained soils cannot.

"Well drained" soils have moderate hydraulic conductivity and seasonal saturation at depths greater than 1.5 m (5 ft). Well drained soils tend to have moisture contents that promote crop growth.

"Moderately well drained" soils have horizons of low hydraulic conductivity and seasonal saturation at depths of 0.90 to 1.5 m (3 to 5 ft). Because of these two factors, these soils have periods of wetness that may limit crop yields or impede planting or harvesting. Fields having moderately drained soils, or worse, may be artificially drained by "field tiles," which are perforated pipes buried in the ground and used to lower the groundwater elevations.

"Somewhat poorly drained" soils experience wetness near enough and long enough to the surface to affect planting and harvesting unless artificial drainage is provided. The soils commonly have horizons of low conductivity and a high water table at some time during the year. Seasonal or periodic saturation is less than 0.3 m (1 ft).

"Poorly drained" soils have extended periods of wetness and may be soils of low hydraulic conductivity. They are evidenced by saturation at or near the ground surface, at 0.3 m (1 ft) or less. These fields must have artificial drainage, or they will be too wet to support crops.

"Very poorly drained" soils are wet to the surface most of the year unless drained and have seasonal saturation at less than 0.3 meters (1 ft).

CONDUCTING A SITE INVESTIGATION

Most onsite wastewater treatment systems rely on native soils to treat and disperse the wastewater from the structures on the premises. If the system is not designed with a full consideration for soil and site limitations, the system might eventually fail. The failure can be troublesome, sometimes causing wastewater backups into the structure, or water pollution and public health nuisances. For these reasons, a major focus of designers and regulators is the soil and site evaluation.

In some jurisdictions, soil and site evaluations are conducted by public health officials. In other jurisdictions, soil and site evaluations may only be performed by licensed personnel.

Soil and site evaluations are more involved than a walk on the site and a quick look at the soil. A complete evaluation includes an understanding of the owner's expectations and knowledge of all the factors that may impact upon the selection and design of a system. Such factors include soil conditions, slope, zoning restrictions, presence of wetlands, setbacks from other structures, wells, property lines, easements, and rights of way.

Meeting the Owner

The first step in conducting a soil and site evaluation is learning about the owner's expectations. The interview should include the following types of questions:

- What is the intended use of the site?
- Are there preferences for building sites?

For a Residence . . .

- How many occupants or bedrooms?
- Intended future land uses for the property.

For a Food service Establishment. . .

- How many patrons or meals will be served, and how many employees?
- Will there be dishwashing?
- Will the food be prepared on the premises?
- What type of food waste will be entering the onsite system?

For a Commercial Building . . .

- What is the intended use of the building?
- Will there be any industrial activities, and how many employees?

The owner should be told about the factors that may influence the location, size, or type of system that may be allowed and how the soil and site limitations may influence the intended location of any structures or roads.

Gathering Background Information

In addition to SCS maps, specific information on a site can be obtained from county extension agents or another agency responsible for onsite systems. These other agencies include the local zoning department or public health agency. Some jurisdictions maintain a record of soil and site evaluations. Public health officials are familiar with local soil conditions and can provide much useful information. A review of local files might show if evaluations of the site or adjoining sites have already been registered.

Deeds, plat records, regulatory agencies, and utilities may provide information on restrictions, designations, and easements that may affect the design. **In all cases where soil borings are contemplated, local utilities should be contacted to ensure that there is no danger of striking underground services.**

Field Investigation

Collecting Information and Tools. Equipped with an understanding of the owner's intentions, expectations, and a review of available information; the site evaluator is ready to conduct the soil and site evaluation.

The first step before going to the site is to sketch out a plot plan, identifying the location and type of all significant features including roads, structures, wetlands, and water bodies. This sketch should be completed on large paper 600 mm X 900 mm (24 in x 36 in), because the sketch will form part of the record for a scaled plot plan.

The second step before going to a site is to assemble the appropriate tools and references necessary to collect the information to complete the plot plan and soil evaluation. Such tools and equipment include but are not limited to:

- Transit
- Level and Level Rod
- Back Hoe
- Munsell Color System book
- Spade or Shovel
- Soil Auger
- Water Bottle
- Soil Description Forms
- Tape Measure
- State or Local Regulations
- Hand Lens
- Putty Knife/Screwdriver/Trowel
- Soil Card
- Dilute Hydrochloric Acid
- Carrying Case
- Tongue Depressors/Nails
- Clip Board
- Reference Soil Samples
- Camera & Film

Preliminary Site Review. Upon arriving at a site, make a tour of site, walking along the property lines. Consider your initial impressions because you will use them to judge the suitability of the site, location and type of system. Look for signs of buried utilities and easements, and confirm that buried utilities are either identified or are absent. In your field notebook note the following observations:

- Topography, including slope direction and grade
- Vegetation type and distribution
- Presence of artificial drainage
- Features on adjacent properties that may affect an onsite system
- Evidence of utilities, such as hydrants power lines, transformers or pedestals
- Location of unidentified structures or features
- Aesthetic features like a tree stand or vista
- Weather conditions during the soil and site evaluation
- Cloud conditions during the soil and site evaluation
- The date time of the soil and site evaluation

Based on your observations and previous discussions, several potential onsite system locations have already been identified. The next step is to perform a preliminary investigation of these sites by using a soil hand auger to examine soil samples. Auguring will either confirm or contradict information provided in a soil survey and provide details not listed in the survey, and this information can be used to select the best potential system location. Identify at least three potential soil boring locations that encompass the proposed disposal site.

Soil borings must be constructed to provide a detailed soil evaluation. Soil borings should be a pit dug to a depth of at least 1.25 meters (4 ft) below the estimated depth of the intended dispersal system. These pits must be wide enough to allow sunlight to enter. Ideally, the borings will be constructed so that the sun shines directly on the exposed face of the profile. Figure 4.8 shows a test hole in Vermont. Borings must also be wide enough to prevent cave-ins or sloughing during the test.

At least one soil boring will need to be constructed, the other two evaluated with a soil auger, if soil conditions are found to be uniform. If soil conditions are not uniform, a soil boring will have to be constructed at each location. Additional borings will be necessary if the soil profiles are markedly different.

DILHR

SOIL AND SITE EVALUATION REPORT

in accord with ILHR 83.05, Wis. Adm. Code

Page 1 of 3

Attach complete site plan on paper not less than 8 1/2 x 11 inches in size. Plan must include, but not limited to vertical and horizontal reference point (BM), direction and % of slope, scale or dimensioned, north arrow, and location and distance to nearest road.

COUNTY: Chippewa
PARCEL I.D. #: 429053541-0000
REVIEWED BY / DATE

APPLICANT INFORMATION–PLEASE PRINT ALL INFORMATION

PROPERTY OWNER: Joseph J. Smith	PROPERTY LOCATION GOVT. LOT NE 1/4 SE 1/4,S 35 T 29 N,R 05 E (or) W
PROPERTY OWNER'S MAILING ADDRESS: Route 1 Box 299	LOT # 12 / BLOCK # 2 / SUBD. NAME OR CSM # Sunnyside Addition
CITY, STATE: Stanley WI / ZIP CODE: 54768 / PHONE NUMBER: (715) 555-1212	[ ] CITY [ ] VILLAGE [X] TOWN: Delmar / NEAREST ROAD: Sunny Lane

[X] New Construction Use [X] Residential / Number of bedrooms 3 [ ] Addition to existing building

[ ] Replacement [ ] Public or commercial describe

Code derived daily flow 450 gpd Recommended design loading rate 0.7 bed, gpd/ft² 0.8 trench, gpd/ft²

Absorption area required 643 bed, ft² 563 trench, ft² Maximum design loading rate 0.7 bed, gpd/ft² 0.8 trench, gpd/ft²

Recommended infiltration surface elevation(s) 92.00 ft (as referred to site plan benchmark)

Additional design / site considerations Trench design recommended.

Parent material Loess over glacial outwash Flood plain elevation, if applicable NA ft

S = Suitable for system / U = Unsuitable for system	CONVENTIONAL	MOUND	IN-GROUND PRESSURE	AT-GRADE	SYSTEM IN FILL	HOLDING TANK
	[X] S [ ] U	[X] S [ ] U	[X] S [ ] U	[X] S [ ] U	[ ] S [X] U	[ ] S [X] U

SOIL DESCRIPTION REPORT

Boring # 1 — Ground elev. 95.00 ft. — Depth to limiting factor > 79"

Horizon	Depth in.	Dominant Color Munsell	Mottles Qu. Sz. Cont. Color	Texture	Structure Gr. Sz. Sh.	Consistence	Boundary	Roots	GPD/ft² Bed	GPD/ft² Trench
Ap	00-10	10YR 3/2	None	SIL	2f-mgr	mfr	as	2f-m	0.5	0.6
E	10-13	10YR 5/3	None	SIL	2mpl	mfr	cs	2f	NP	0.2
E/B	13-16	10YR 5/3 & 10YR 4/4	None	SIL	2mpl	mfr	cw	1f	NP	0.2
Bt1	16-25	10YR 4/4*	None	SIL	2msbk	mfr	cw	1m	0.5	0.6
Bt2	25-28	7.5YR 5/4*	None	SIL	2msbk	mfr	cw	-	0.5	0.6
2Bt3	28-32	5YR 4/4	None	SL	2msbk	mfr	as	-	0.5	0.6
2C	32-79	7.5YR 5/4	None	S	0sg	ml	-	-	0.7	0.8

Remarks: * Common 10YR 7/2 silt coatings on ped surfaces.

Boring # 2 — Ground elev. 96.00 ft. — Depth to limiting factor > 85"

Horizon	Depth in.	Dominant Color Munsell	Mottles Qu. Sz. Cont. Color	Texture	Structure Gr. Sz. Sh.	Consistence	Boundary	Roots	GPD/ft² Bed	GPD/ft² Trench
1	00-09	10YR 3/2	None	SIL	2mgr	mfr	as	2f-m	0.5	0.6
2	09-14	10YR 5/3 & 10YR 4/4	None	SIL	2msbk	mfr	cw	1f	0.5	0.6
3	14-22	10YR 4/4*	None	SIL	2msbk	mfr	cw	-	0.5	0.6
4	22-32	7.5YR 5/4*	None	SL	2msbk	mfr	as	-	0.5	0.6
5	32-85	7.5YR 5/4	None	S	0sg	ml	-	-	0.7	0.8

Remarks: * Common 10YR 7/2 silt coatings on ped surfaces.

CST Name:—Please Print: Dusty Doe Phone: (715) 555-2121

Address: 123 First Ave., Chippewa Falls, WI 54729

Signature: Dusty Doe Date: June 21, 1992 CST Number: M0000

Figure 4.7
Soil & Site Evaluation Report

Conducting Detailed Soil Evaluations

Once a soil boring has been constructed and is secure for entry[b], the site evaluator should make general observations from the surface. All observations must be recorded on the Soil & Site Evaluation Report, an example of which is shown in Figure 4.7. Is there anything unusual about the profile? Is water seeping in and, if so, from where? Are the horizons generally parallel to the ground surface? Are they wavy? Distinct? How does the overall impression of the profile compare with expectations based on surface topography and vegetation? These observations should be recorded and then the boring should be entered for a detailed evaluation of the soils.

Figure 4.8 - Test hole in Vermont

Horizon Identification and Marking. First, identify each horizon and mark it, if necessary, with tongue depressors. Marking horizons makes depth measurements easier and helps to organize the horizon-by-horizon evaluation. Starting with the surface horizon and working to the bottom of the boring, list each horizon on your evaluation form as #1, #2, #3, and so forth. For general purposes, it is not necessary to identify a horizon by its SCS designation since the purpose of the evaluation is solely to judge suitability for an onsite system. From the ground surface, measure the depth to each horizon. Once you have identified each horizon and measured its depth, you are ready to evaluate each horizon using the same evaluation routine, recording your observations as you make them and before proceeding to the next horizon.

Soil Texturing. Soil texture refers to the composition of the soil based on the percentages of sand, silt, or clay as shown in the USDA Soil Triangle shown in Figure 4.2 on Page 62. Soil texture greatly affects the decisions about siting and design of an onsite system. Consequently, texturing is used to classify the soil and thus judge its suitability.

---

[b]Excavations are subject to the requirements of OSHA. Please contact the nearest office of the Department of Labor, Occupational Safety and Health Administration and request a copy of "Safe Practices for Excavation and Trenching Operations," which details the safety requirements for entering and working in excavations.

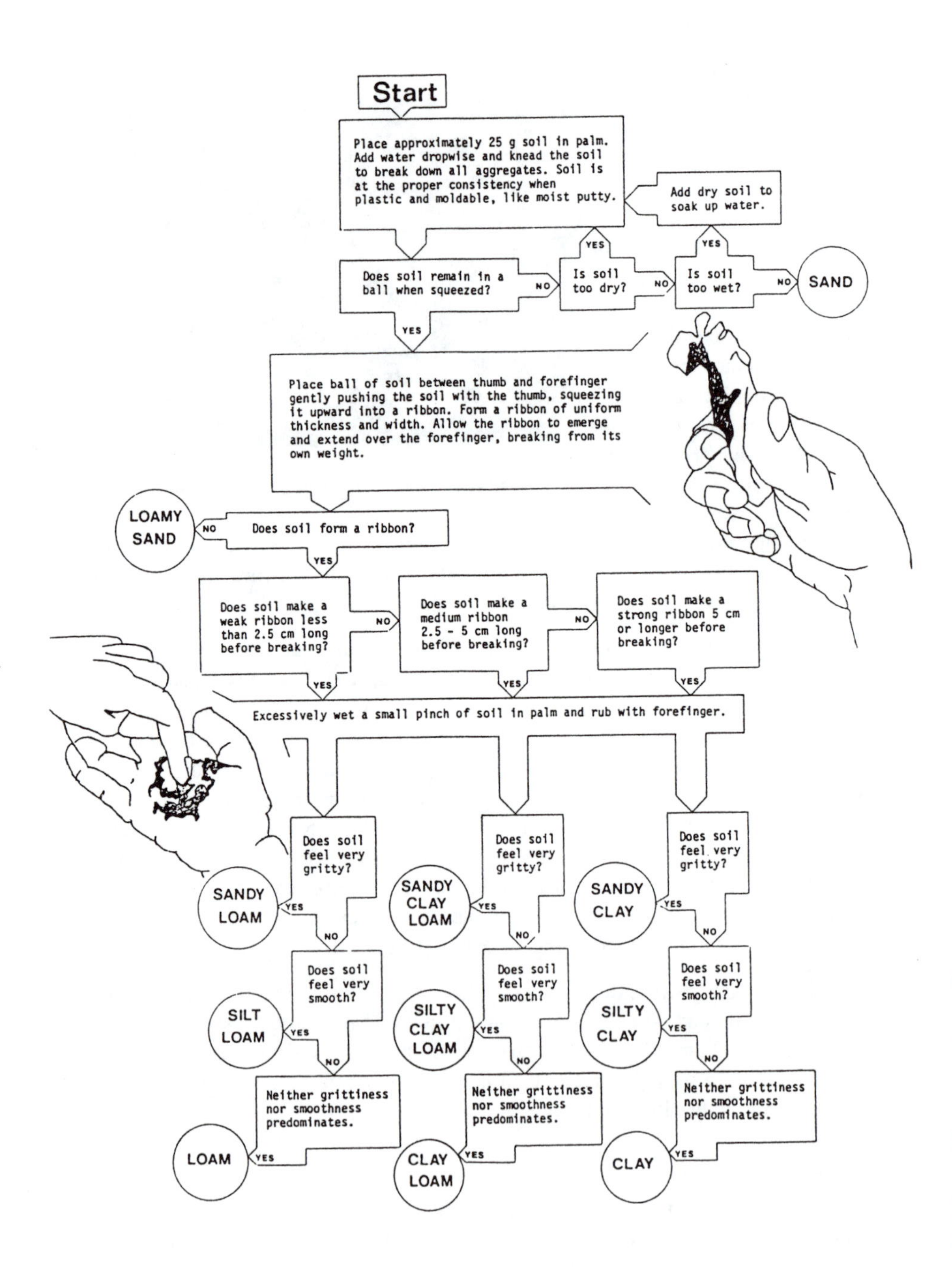

Figure 4.9 - Steps in soil texturing

Soil texturing in a laboratory is conducted through a sieve and hydrometer or pipette analysis. Such analyses are expensive and time-consuming. However, fairly reliable, empirical texturing can be done in the field.

Because the fine earth fraction consists of tiny soil particles touch — not sight — is used to texture the soil. Different types of soils will exhibit different properties, and similar soils will exhibit similar properties. The steps in texturing a soil are shown in Figure 4.9. It is a skill that can be developed with a little practice. Some points to remember in texturing soils are as follows:

- Wet the soil so that it simulates normal field conditions. A dry soil acts different from a wet one. Wet the soil until it is plastic.

- Sand grains and coarse particles are easily felt, and sandy soils do not hold together as easily as finer-texture soils. Fine sand may not be easily felt.

- Moist silt loams can be squeezed together but cannot be mashed easily into a ribbons, which will be less than 25 mm (1 in) long. The soil feels smooth.

- Clay loams and clays will easily form ribbons. Clay loam ribbons are 25 mm to 50 mm long (1 in to 2 in), Clays form ribbons over 50 mm (2 in) long. Neither grittiness nor smoothness prevails.

Using the Munsell Color System, identify the colors that dominate the horizon. If necessary, interpolate between colors and feel free to examine multiple samples. Use the best light and no tinted eyeglasses to evaluate the colors. Leave the boring, if necessary, to get better light. Examine the soil for mottles (redoximorphic features) both in the samples and by digging into the profile with a putty knife or other tool. Record the quantity, color, and size of any mottles that are present. Be certain to note the depth of mottles that might be indicative of seasonal saturation.

Examine the soil structure of the sample before you texture it. What are the characteristics of the peds? What are their grade, type, and size? Texture soil samples with your hand. Touch, rub, and examine the soil, taking samples in hand and by cutting into the horizon. Can the soil be rubbed into a rope? How moist is it? How does the soil compare with identified reference samples? Identify the soil according to one of the twelve SCS classifications. Before you leave the horizon, examine its face and dig into it to establish the quantity and size of any roots and the distinctness of the boundary between adjacent horizons. Record your observations before proceeding to the next horizon.

When you have completed your evaluation, review and check it again. Take out some more soil samples and compare them with your initial evaluation. Are your second observations consistent with your first? If they are, then you are ready to close the boring. If they are not, examine why you have changed your mind, examine more soil samples, and make final corrections as appropriate. Close the boring, or if it must remain open for regulatory agency review, protect it from damage or unauthorized entry.

Confirm the soil profile with augured soil samples from the other identified locations, and if confirmation cannot be obtained, then construct additional soil borings and perform soil evaluations until the site can be accurately characterized. Show restraint in the number and location of soil borings. Too many soil borings can destroy an otherwise acceptable site.

Site Survey. After the detailed soil evaluation is completed, the next step is to conduct a site survey using the survey instruments. The first step in conducting the survey is to identify a permanent benchmark for referencing elevations and a permanent reference line for measuring angles and distances. After the benchmark and baseline have been identified, the location and elevation of significant features can be identified.

- First, the location of all soil borings must be identified.
- Second, the ground elevation of the soil boring locations and surrounding area must be established.
- Third, the slope of the surrounding area must established so that elevation contours can be identified.
- Fourth the location and elevation of water bodies, structures, and other significant items listed.

Soil Permeability

Some designers are most comfortable designing soil absorption systems based on the measured value of soil permeability. The standard test for soil permeability is detailed in ASTM D2434 and may be conducted by one of two methods,the *constant head* test or the *falling head* test. The principle behind both tests is identical: apply Darcy's Law (p.89), based on the measurements taken, to calculate the coefficient of permeability, k, for an **undisturbed** soil sample.

In the constant head test, an undisturbed sample is placed in the apparatus and subjected to an upward flow - at a constant head - through the bottom of the sample. k is calculated my measuring the head loss, flow, and thickness of the sample. In the falling head test, k is calculated by measuring the drop in head and corresponding, again applying Darcy's Law. Sample apparatus to measure constant head is shown is Figure 4.10.

Soil permeability tests have significant limitations. Their use is limited to homogeneous soils. And, the test actually measures flow in only in the direction of the vertical flow. These limitations can pose a problem for two reasons: most soils are not homogeneous, and water flows through soil both horizontally and vertically. There are techniques for estimating the permeability of non-homgenous soils, but such techniques are beyond the scope of this text. Likewise, the consideration of two dimensional flow is beyond the scope of this text. Despite these inherent limitations, permeability tests provide information useful to the designer.

There are methods to measure permeability in the field, one of which is well known to well drillers. This method is a "pumping test." The test consists of drilling a hole into an aquifer whose soil characteristics are known and measuring the flow from the aquifer. The test is conducted by pumping water from the aquifer at a steady rate and calculating the permeability using standard formulae for that purpose. This approach is acceptable only if the soils in the

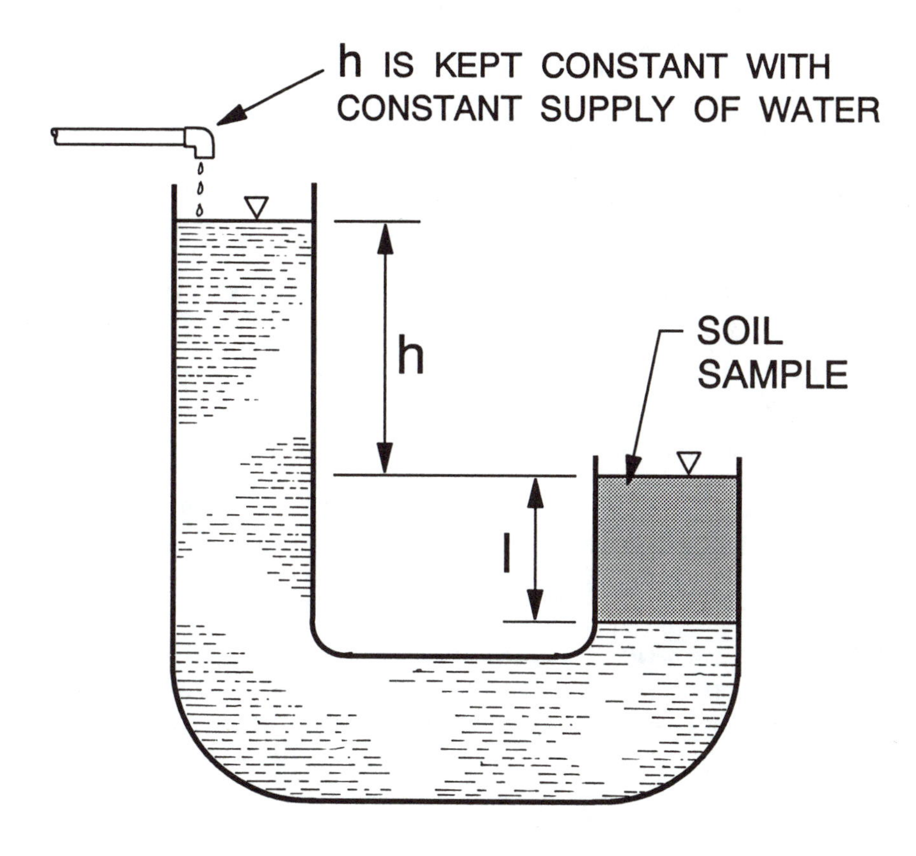

Figure 4.10 - Apparatus for a Constant Head Test

aquifer were identical to the soils above it. Alternatively, permeability can be estimated by augering a hole into an aquifer and measuring the rate at which it fills with water. As with the pumping test, standard formulae are used to relate the test data to soil permeability.

Once all of these measurements are made, the site evaluator is ready to return to the office tó prepare a detailed plot plan and make determinations of the suitability of the site. The plot plan should be laid out on a sheet at a scale of no less that 1:500 (1 inch equals 40 feet). Larger scales are preferable. At minimum, the plot plan must include the following items for the site under evaluation and for adjacent properties if the features might influence the siting or selection of an onsite system:

- Date
- Name of Site Evaluator
- Business Address
- License Number
- Site Location
- Site Legal Description
- Scale of Drawing
- Soil Borings
- Existing Structures
- 2-foot Contours of Site
- Location of Wells
- Easements
- Property Boundaries
- Weather Conditions
- Benchmark
- Water bodies
- Time of Evaluation
- Unusual Conditions, such as drought

The completed plot plan and soil evaluation forms should be included as part of a report that summarizes the findings and recommendations of the site evaluator. The plot plan should identify:

- Site limitations, such as excessive slope
- Restrictions, such as setbacks from structures
- Easements and rights of way
- Prominent aesthetic features

The report should detail any soil and site limitations such as high groundwater, bedrock, or impermeable soils. It might also suggest suitable system alternatives and the suggested geometry for a soil absorption system.

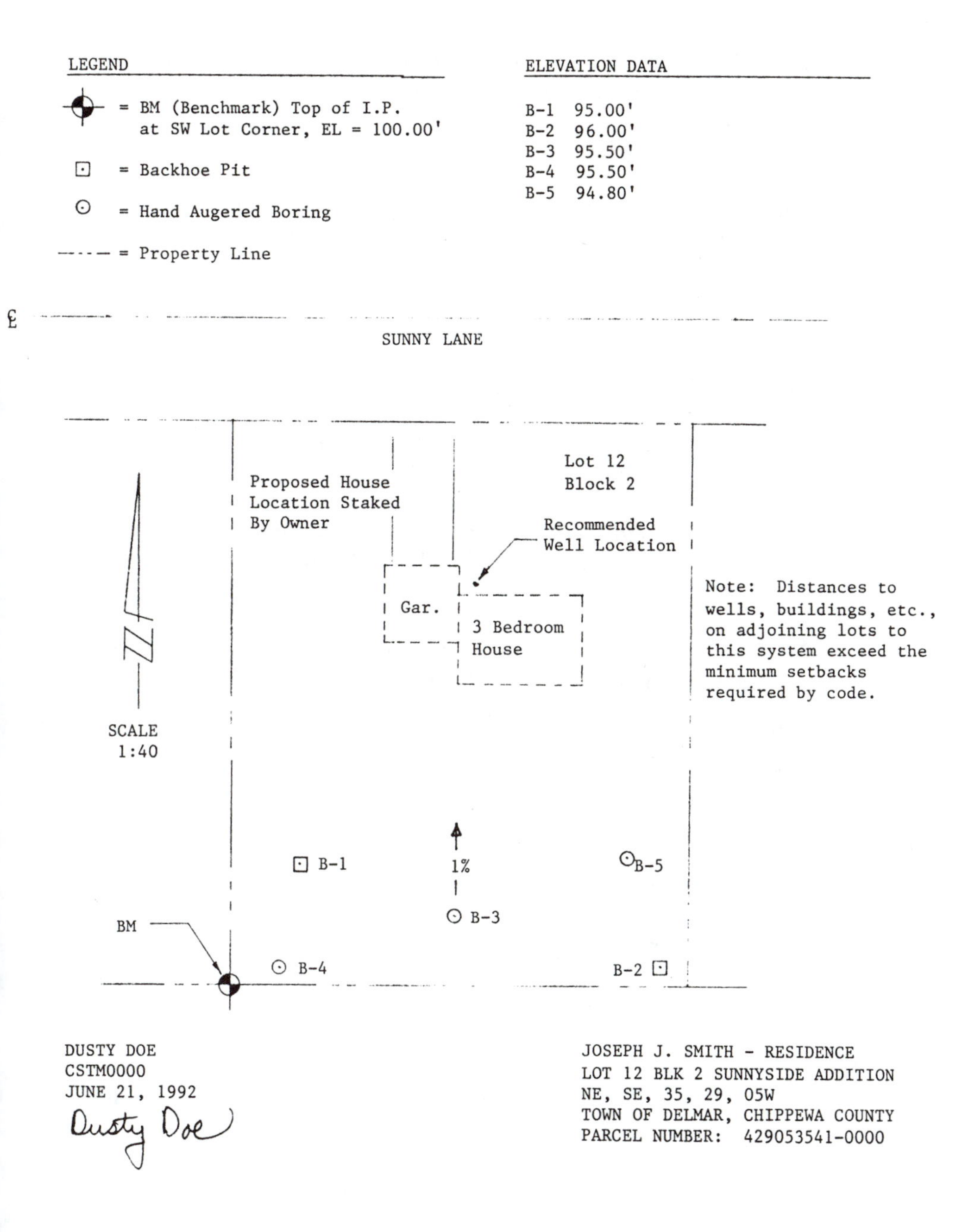

Figure 4.11 - Sample Plot Plan

References

1. Hausenbuiller, R. *Soil Science Principles and Practices*. Dubuque: Wm. C. Brown, Company Publishers, 1972, pp 169-187.

2. Hausenbuiller. pp 45-53

3. Dunn, I.; Anderson; L.; Kiefer, F. *Fundamentals of Geotechnical Analysis*. New York: John Wiley & Sons, 1974, pp 17-19.

4. Brady, Nyle C. *The Nature and Properties of Soils*, 8th Edition. Macmillan Publishing Company, inc. New York, 1974. p. 259.

5. Dunn, et. al. p 14.

6. Buol, S; Hole, F.; McFracken, R. *Soil Genesis and Classification*. Ames: The Iowa State University Press, 1973, pp 88-100.

7. North Carolina Agricultural Research Service. *Redoximorphic Features for Identifying Aquic Conditions*, Technical Bulletin 301. Raleigh: North Carolina State University, 1992.

8. Wisconsin Department of Commerce (formerly Wisconsin Department of Industry, Labor and Human Relations) *Soil and Site Evaluation Handbook*, SBD-9046-P (R.02/93), Madison, WI, pp 41, 43.

9. Dunn, et.al. p 49

10. Hausenbuiller, pp 106-111

11. Wisconsin Department of Commerce, p 39.

12. Wisconsin Department of Commerce, p 61.

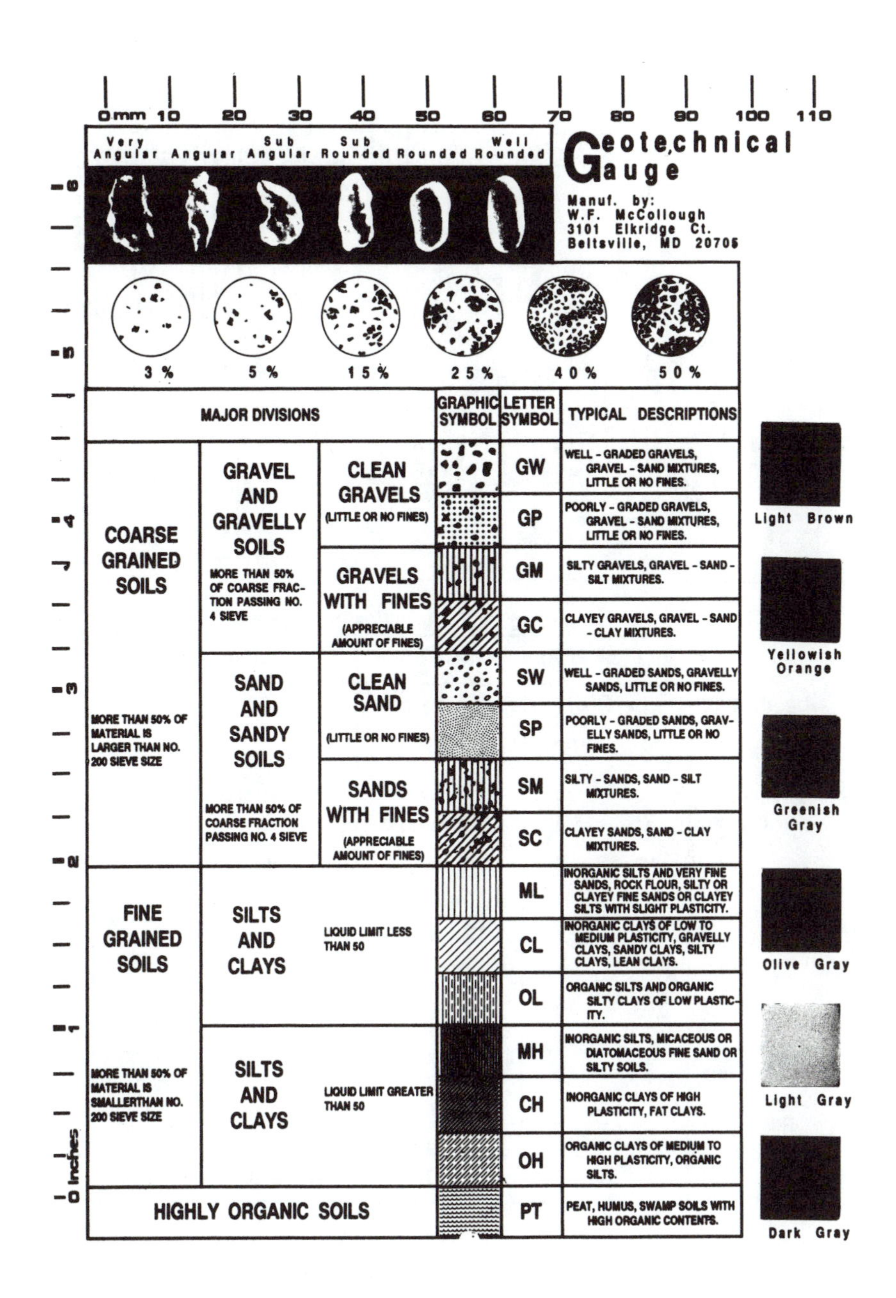

MAJOR DIVISIONS			GRAPHIC SYMBOL	LETTER SYMBOL	TYPICAL DESCRIPTIONS
COARSE GRAINED SOILS MORE THAN 50% OF MATERIAL IS LARGER THAN NO. 200 SIEVE SIZE	GRAVEL AND GRAVELLY SOILS MORE THAN 50% OF COARSE FRACTION PASSING NO. 4 SIEVE	CLEAN GRAVELS (LITTLE OR NO FINES)		GW	WELL - GRADED GRAVELS, GRAVEL - SAND MIXTURES, LITTLE OR NO FINES.
				GP	POORLY - GRADED GRAVELS, GRAVEL - SAND MIXTURES, LITTLE OR NO FINES.
		GRAVELS WITH FINES (APPRECIABLE AMOUNT OF FINES)		GM	SILTY GRAVELS, GRAVEL - SAND - SILT MIXTURES.
				GC	CLAYEY GRAVELS, GRAVEL - SAND - CLAY MIXTURES.
	SAND AND SANDY SOILS MORE THAN 50% OF COARSE FRACTION PASSING NO. 4 SIEVE	CLEAN SAND (LITTLE OR NO FINES)		SW	WELL - GRADED SANDS, GRAVELLY SANDS, LITTLE OR NO FINES.
				SP	POORLY - GRADED SANDS, GRAVELLY SANDS, LITTLE OR NO FINES.
		SANDS WITH FINES (APPRECIABLE AMOUNT OF FINES)		SM	SILTY - SANDS, SAND - SILT MIXTURES.
				SC	CLAYEY SANDS, SAND - CLAY MIXTURES.
FINE GRAINED SOILS MORE THAN 50% OF MATERIAL IS SMALLERTHAN NO. 200 SIEVE SIZE	SILTS AND CLAYS	LIQUID LIMIT LESS THAN 50		ML	INORGANIC SILTS AND VERY FINE SANDS, ROCK FLOUR, SILTY OR CLAYEY FINE SANDS OR CLAYEY SILTS WITH SLIGHT PLASTICITY.
				CL	INORGANIC CLAYS OF LOW TO MEDIUM PLASTICITY, GRAVELLY CLAYS, SANDY CLAYS, SILTY CLAYS, LEAN CLAYS.
				OL	ORGANIC SILTS AND ORGANIC SILTY CLAYS OF LOW PLASTICITY.
	SILTS AND CLAYS	LIQUID LIMIT GREATER THAN 50		MH	INORGANIC SILTS, MICACEOUS OR DIATOMACEOUS FINE SAND OR SILTY SOILS.
				CH	INORGANIC CLAYS OF HIGH PLASTICITY, FAT CLAYS.
				OH	ORGANIC CLAYS OF MEDIUM TO HIGH PLASTICITY, ORGANIC SILTS.
HIGHLY ORGANIC SOILS				PT	PEAT, HUMUS, SWAMP SOILS WITH HIGH ORGANIC CONTENTS.

Geotechnical Gauge - Side 1

(Available from W.F. McCollough, 3101 Elkridge Ct., Beltsville, MD 20705)

CLAY

CLAY CONSISTENCY	THUMB PENETRATION	SPT, N BLOWS/FT.	Undrained Shear Strength c (PSF) TORVANE	Unconfined Compressive Strength q_u Pocket Penetrometer
VERY SOFT	Easily penetrated several inches by thumb. Exudes between thumb and finger's when squeezed in hand.	< 2	250	500
SOFT	Easily penetrated one inch by thumb. Molded by light finger pressure.	2 - 4	250 - 500	500 - 1000
MEDIUM STIFF	Can be penetrated over 1/4 " by thumb with moderate effort. Molded by strong finger pressure.	4 - 8	500 - 1000	1000 - 2000
STIFF	Indented about 1/4" by thumb but penetrated only with great effort.	8 - 15	1000 - 2000	2000 - 4000
VERY STIFF	Readily indented by thumbnail.	15 - 30	2000 - 4000	4000 - 8000
HARD	Indented with difficulty by thumbnail.	> 30	> 4000	> 8000

SAND

SOILTYPE	SPT, N Blows/ft.	Relative Density,%	FIELD TEST
VERY LOOSE SAND	4	0 - 15	Easily penetrated with 1/2" reinforcing rod pushed by hand.
LOOSE SAND	4 - 10	15 - 35	Easily penetrated with 1/2" reinforcing rod pushed by hand.
MEDIUM DENSE SAND	10 - 30	35 - 65	Penetrated a foot with 1/2" reinforcing rod driven with 5-lb hammer.
DENSE SAND	30 - 50	65 - 85	Penetrated a foot with 1/2" reinforcing rod driven with 5-lb hammer.
VERY DENSE SAND	50	85 - 100	Penetrated only a few inches with 1/2" reinforcing rod driven with 5-lb hammer.

Unified Soil Classification System (USCS)

	MILLIMETERS	INCHES	SIEVE SIZES
BOULDERS	> 300	> 11.8	-
COBBLES	75 - 300	2.9 - 11.8	-
GRAVEL: COARSE	75 - 19	2.9 - .75	-
FINE	19 - 4.8	.75 - .19	3/4" - No. 4
SAND: COARSE	4.8 - 2.0	.19 - .08	No. 4 - No. 10
MEDIUM	2.0 - .43	.08 - .02	No. 10 - No. 40
FINE	.43 - .08	.02 - .003	No. 40 - No. 200
FINES: SILTS	< .08	< .003	< No. 200
CLAYS	< .08	< .003	< No. 200

Geotechnical Gauge - Side 2
(Actual sand particles glued to laminated card)

CHAPTER 5

HYDRAULICS

Topics of This Chapter:

- *Hydraulic Theory*
- *Design Concepts in Practice*

FUNDAMENTAL CONCEPTS

Understanding the movement of wastewater in onsite wastewater treatment systems is a major requirement of a designer. Water moves from the structure through pipes into a septic tank, a grease trap, a treatment unit, or an equalization basin. The wastewater remains in this vessel for a period established during the design. The effluent may be pumped or flow by gravity into either another treatment unit or directly into pipes that convey the effluent to the soil absorption area. There, the effluent is dispersed into soil. It is imperative to understand how water (often with many things dissolved or suspended in it) behaves under each of these conditions. The behavior of water under each of these circumstances can be explained by a study of *hydraulics*. This chapter is important to designers, but may be skipped by the more casual reader without losing the continuity of the book

Hydraulics is the study of the behavior of liquids at rest and in motion. Recall that matter can exist in one of three states: solid, liquid or gas. Gases and liquids are both considered *fluids*. Gases can be compressed. A gas will fill the space available for it. Liquids are considered incompressible and have a definite volume and a surface layer. All fluids are unable to resist *shearing stresses*, those stresses that separate the molecules binding the fluid together. Because fluids cannot resist shear, they *flow* and transmit the stresses to all solid boundaries or any surface in contact with the fluid. Hydraulics helps us understand the effects of flow and the transmission of stresses[1].

Flow

Liquids exhibit two kinds of flow: laminar and turbulent. Laminar flow is characterized by a lack of mixing among the molecules; the flow proceeds in an orderly pattern, as shown in Figure 5.1a. When a fluid at rest starts to move, it must overcome resistance to flow due to the viscosity of the fluid. Slowly moving fluids are dominated by viscous forces. Energy levels are low, and the resulting fluid flow is laminar. In laminar flow, molecules of water move in smooth lines, called streamlines. Water in saturated soil moves by laminar flow, and the velocities can be measured in distance of movement per year (either meters/year or feet/year). As the velocity increases, the fluid gains kinetic energy. Eventually, the inertial forces due to movement are more influential than the viscous forces, and the fluid particles begin to rush past each other in an erratic fashion. The result is turbulent flow, in which the water molecules no longer move along parallel streamlines. Turbulent flows are generally seen in "whitewater" streams and found in piping systems, as is shown in Figure 5.1b[2].

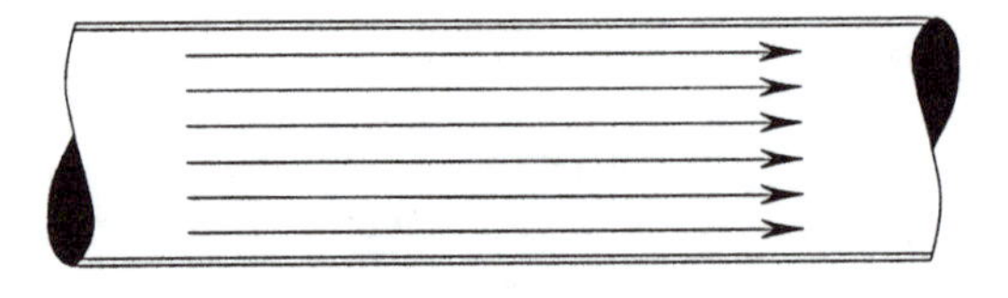

5.1*a* Laminar Flow

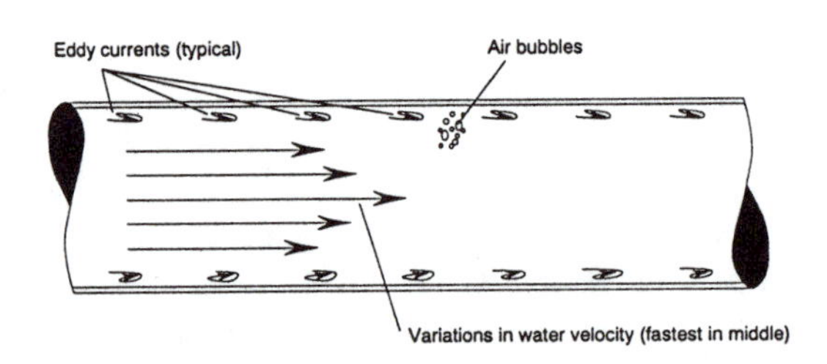

5.1*b* Turbulent Flow

Figure 5.1*a* Laminar Flow and 5.1*b* Turbulent Flow

Around 1856, a French hydraulic engineer, Henry Darcy, made the first systematic study of the movement of water through a porous medium. He studied the movement of water through beds of sand used for water filtration. Darcy found that the rate of water flow through a bed of sand was proportional to the difference in the height of the water between the two ends of the filter beds and inversely proportional to the length of the flow path. He also determined that the quantity of flow is proportional to a coefficient, K, which is dependent upon the nature of the porous medium.

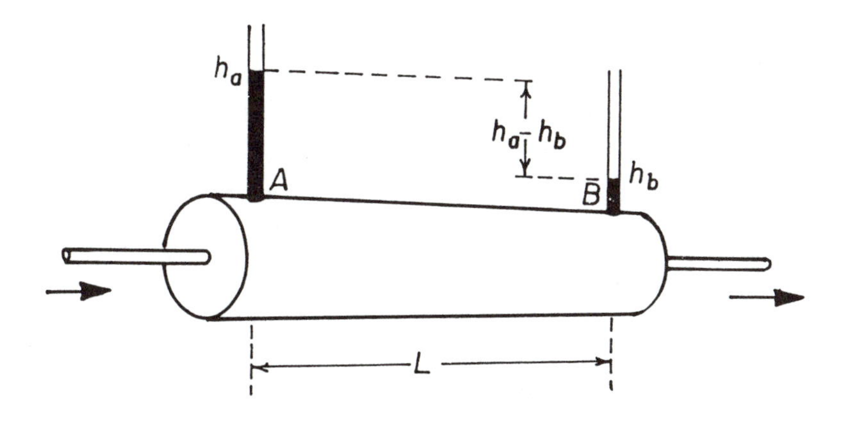

Figure 5.2 - Darcy's Law Illustration

Figure 5.2 illustrates a tube of cross-section *M* filled with sand. Water is applied under pressure through one end, A, until all the pores between the sand particles are filled with water and the inflow rate equals the outflow rate. The pressure at each end can be measured and observed by means of a thin vertical pipe in the sand at each end. The distance between the two vertical pipes (which are pizeometers or pressure meters) is Δl.

Darcy demonstrated that the discharge, Q, is proportional to the difference in the height of the water, h (hydraulic head), between the two end tubes and inversely promotional to the flow length, L: $Q \propto h_a - h_b$ and $Q \propto 1/L$. Hydraulic head is expressed as feet or meters of head.

Flow is also proportional to the cross-sectional area, A, of the pipe. When combined with the proportionality constant, K, the result is the expression known as Darcy's Law:

$$Q = KA\left(\frac{h_a - h_b}{L}\right) \quad \textbf{(5.1)}$$

This may be expressed in more general terms as

$$Q = -KA\left(\frac{dh}{dl}\right) \tag{5.2}$$

where
dh/dl is known as the *hydraulic gradient.*

The quantity *dh* represents the change in hydraulic head between two points that are very close together, and *dl* is the small distance between these two points. The negative sign in Equation 5.2 indicates that flow is always in the direction of decreasing hydraulic head. The use of the negative sign necessitates careful determination of the sign of the gradient. If the value of h_2 at point X_2 is greater than h_1 at point X_1, then flow is from point X_2 to X_1. If $h_1 > h_2$, then flow is from X_1 to X_2. Under most natural groundwater conditions, the velocity is sufficiently low for Darcy's law to be valid[3].

The last equation can be rearranged to show that the coefficient *K* has the dimensions of length/time (*L/T*), or velocity. This coefficient has been termed the *hydraulic conductivity* and has high values for sand and gravel and low values for clay and most rock. Examples of the units that are used for values of *K* are m/day (ft/day).

Surface Tension

All liquids exhibit tension at their surfaces. The *surface tension* of water is described in Chapter 2, *Water and Wastewater Characteristics*. Surface tension, T_s (or σ) — measured as N/m (ft/lb) — is the condition that results when the surface of a liquid acts like a membrane that is able to support a load. Surface tension is the force on the water surface perpendicular to a line drawn along the surface. Water has a surface tension of 0.0728 N/m at 20°C, as compared to other liquids that have surface tensions ranging from 0.020 — 0.040 N/m. Figure 5.3 shows how the surface tension force is determined using a glass tube. Surface tension is also exhibited wherever two fluids are in contact with each other or wherever a liquid is in contact with a solid. The strength of the tension depends upon the relative sizes of the attractive forces between the two liquids

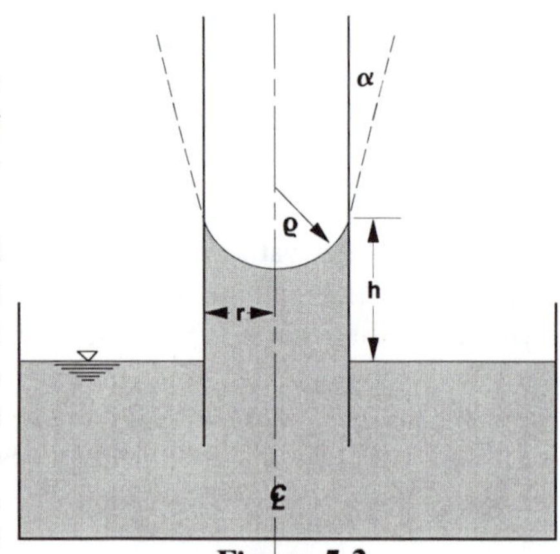

Figure 5.3
Surface Tension in a Glass Tube

or the liquid and the solid. For many engineering problems involving unit operations, the effects of surface tension are negligible[4].

$$T_s = \frac{rhdg_n}{2} \tag{5.3}$$

where

T_s = surface tension, N/m (lb/ft)
r = internal radius of capillary tube, m (ft)
h = height water in soil rises, m (ft)
d = density of the fluid, kg/m^3 (lb/ft^3)
g_n = gravitation acceleration, 9.807 m/sec^2 (32.174 ft/sec^2)

In a laboratory, capillary action is easily demonstrated by observing how high water rises inside fine-diameter tubes. Experiments can show how high this force will permit the water to rise in tubes of different diameters, as you saw on page 68. Many of the relationships developed for glass tubes can be directly applied to soils. To fully describe the behavior and movement of water in soil, surface tension must be considered because it is a major factor. In soil, the high surface tension of water interacting with the tiny pore spaces among the soil particles results in *capillary action*[5].

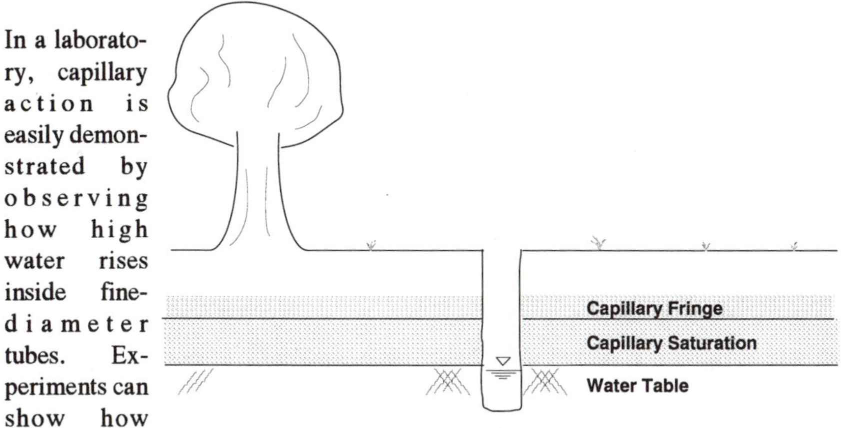

Figure 5.4 - Capillary Action in Soils

Equation 5.4 shows the relationship among the many physical factors in the soil/water matrix, as modelled by a glass tube. This equation shows the height of the water rise is mostly a function of the radius of the tube. From this equation, the height of the water is *directly* proportional to the specific gravity of the fluid and *inversely* proportional to the diameter of the tube; in the tiniest tubes, water rises the highest. Relating glass tubes to soils, the finer the particles, the more tightly the water will be bound and the higher it will rise in the soil matrix, as shown in Figure 5.4.

$$h = \frac{2T_s}{\gamma_w r} \cos\alpha \qquad (5.4)$$

where

h = height of water in a tube, meters
T_s = surface tension of water, 0.0735 N/m
r = radius of the tube or pore, meters
γ = specific weight of water, 9,807 N/m^3
α = angle of meniscus, degrees
ρ = radius of the meniscus, degrees

In soil, water rises through capillary action, creating a zone known as the *capillary fringe*. In this fringe, capillary action exerts a significant stabilizing force on the soil particles. The magnitude of this force at the top of the capillary zone can be calculated and is called capillary tension. The pressure difference between an air-water interface is calculated thus:

$$u_b = \frac{-2T_s}{r} \cos\alpha \qquad (5.5)$$

where

u_b = pressure difference, kN/m^2 (For a glass tube, α = 0 and ρ = r)

Table 5.1 lists the average capillary heights and tensions in soils[6].

Table 5.1
Capillary Heights and Tension for Various Soils

Material	Height (m)	Tension (KPa)
Sand	0.05 - 1	0.5 - 10
Silt	1 - 10	10 - 100
Clay	> 10	>100
Maximum	>35	>350

Energies in a System

In an ideal fluid — one that experiences no friction losses — the total energy remains constant. This phenomenon is known as the *Bernoulli Constant*; from this it follows that the total energy in a hydraulic system is the sum of the energy exerted by forces on the surface of the water, the velocity of the water, and the surface elevations above a datum (or baseline) at the start and end of the system. This energy is expressed for two random locations, 1 and 2, below:

$$\frac{p_1}{\gamma} + \frac{V_1^2}{2g_n} + z_1 = \frac{p_2}{\gamma} + \frac{V_2^2}{2g_n} + z_2 \qquad (5.6)$$

where

p = pressure at points 1 and 2, N/m^2 (lb/ft^2)
γ = Specific gravity, N/m^3, (lb/ft^3)
V = velocity, m/sec (ft/sec)
g_n = gravitational acceleration, 9.807 m/sec^2 (32.174 ft/sec^2)
z = elevations above a datum or baseline, m (ft)

The *pressure head* is the sum of the atmospheric pressure exerted on the surface of a liquid plus the weight of any other fluids or solids pressing upon a liquid. The *velocity head* is the movement — if any — of a liquid, and the *elevation head* of a liquid is its height of the liquid surface above a baseline. The Bernoulli equation is expressed in units of length, *meters (feet) of head*, that is, the pressure at the bottom of a water column. Head is used because all liquids can be envisioned as existing in a column that exerts a pressure perpendicular against all submerged objects[7].

Water is the only liquid that is considered in this text. The nominal specific weight of water at 20°C (65°F) is 1 kN/m^3, γ= 1 kN/m^3 = 1 kPa = 1 m head. (62.34 lbs/ft^3 for U.S. customary units where 1 lb/in^2 equals 144 lbs/ft^2. Therefore, a column of water 2.31 feet high (2.31 feet of "head") produces a pressure equal to 1 lb/in^2).

Ideal fluids are a convenient concept for the derivation of the formulae that explain important hydraulic relationships. In practical applications, the designer must always deal with friction, the force that robs the system of its energy. Friction results from many sources within the system. The flow itself will be turbulent. As a result, energy will be lost in the swirling of the water. The pipe will not be perfectly smooth, so the pipe will warm as the water passes through it. Fittings will force the water to change direction or to squeeze through constrictions, losing more energy with each maneuver. Valves will reduce the cross-sectional area, resulting in even more frictional losses.

For engineering determinations, the Bernoulli equation can be written to account for the losses thus:

$$\frac{p_1}{\gamma} + \frac{v_1^2}{2g_n} + z_1 = \frac{p_2}{\gamma} + \frac{v_2^2}{2g_n} + z_2 + h_L \qquad (5.7)$$

where h_L is proportional to the *flow* in the fitting, valve, or other component.

Flow — the product of the water velocity and cross-sectional pipe area — can be used to estimate the size of the headloss. Headlosses are calculated in terms of the velocity head of the system:

$$h_L = K_L \frac{V^2}{2g_n} \quad (5.8)$$

where
K_L equals loss coefficient.
Loss coefficients for common fittings are shown in Table 5.2[8].

Table 5.2
K_L for Pipe Fittings

Fitting	1-inch	1½-inch	2-inch
Gate Valve	0.18	0.15	0.15
Ball Valve	0.07	0.06	0.06
Plug Valve	0.41	0.38	0.34
Elbow,45°	0.37	0.63	0.57
Elbow, 90°	0.69	0.63	0.57
Tee, thru	0.46	0.42	0.38
Tee, branch	1.38	1.26	1.14

Friction losses in lengths of straight pressurized pipe can be calculated using the Hazen-Williams Formula, Equation 5.10. The Hazen-Williams coefficient for common materials is shown in Table 5.3.

An abrupt change of pipe diameter is a special case of the general principle. In his case, the headloss can be calculated as follows:

$$h_L = \frac{(V_1 - V_2)^2}{2g_n} \quad (5.9)$$

where $(V_1\text{-}V_2)$ equals the change in flow as a result of changes in pipe diameters.

Table 5.3
Hazen-Williams Coefficients

Type of Pipe Equipment	New	Used
Plastic	150	140
Cement Lined	150	140
Copper	140	130
Cast Iron	130	100
Concrete	120	100
Steel	100	90

To calculate the headloss, the formula is rewritten to solve for the slope, the headloss per foot of length. The headloss in the piping system is the headloss per foot times the total length of pipe. For convenience, this equation has been rewritten and converted to U.S.Customary units for practical design, as shown in Equation (5.10):

$$V = K_{hw} C_{hw} R_h^{0.63} S^{0.54} \tag{5.10}$$

where

V = Velocity, ft/sec
K_{hw} = Constant (1.318 for US Customary Units, 0.849 for SI Units)
C_{hw} = Hazen-Williams Coefficient
R_h = Hydraulic Radius, m (ft)
S = Slope, drop/run, m/m (ft/ft)

Equation (5.10) can be written to calculate headloss using units in common practice:

$$h_L = 0.002083 L \left(\frac{100}{C_{hw}}\right)^{1.85} \left(\frac{Q_{gpm}^{1.85}}{d^{4.8655}}\right) \tag{5.11}$$

where

Q_{gpm} = flow, gal/min
d = pipe diameter, in.
L = Pipe length, ft.

Friction losses also occur at points of entrance and discharge from the system. The magnitude of the loss varies with the type of orifice for entrance or discharge. K_L for entrance losses varies from 0.04 for large, well-rounded orifices to 0.50 for sharp-edged orifices. Inward projecting orifices have a K_L of 0.78. K_L for discharge points is 1.00[9].

System Curves

A *system curve* is method of selecting the proper pump for a distribution system. The curve is developed by calculating and plotting the total headloss for a distribution system. On the same plot, the *pump curve* for specific pumps will be drawn. The expected actual flow will be the intersection of the pump curve and the headloss curve.

The system curve is plotted, flow versus total head. This selection provides a convenient way of observing how a change in head or an increase in headloss affects the operation of the distribution system[10].

A pumping and distribution system will function properly only if the correct pump is selected, and the correct pump can be selected only if the Bernoulli Equation variables are known. For an onsite system, the critical number is the total headloss because it is always the most difficult factor to estimate. The pressure at both the intake and discharge is atmospheric pressure, so the pressure variable drops out. The flow, and thus the velocity of the water, will decrease as the length of piping and number of fittings increase. The elevation at both locations can be easily measured.

Total headloss is the sum of the headloss coefficients for all piping, fittings, valves, and intake and discharge points. Each of these is directly proportional to the condition of the piping, the appurtenances and the flow. Headlosses increase as a system ages and the mechanical components begin to deteriorate. Figure 5.5 shows a typical system curve for a simple pumping system. The system curve is developed by first establishing the flow required at the outlet. The components needed to convey the wastewater to the discharge point are then designed. Headlosses are then calculated for the system when it is new and for a time when the piping and appurtenances have aged, usually a 20-year period[11].

The next step is to identify proper pumps in a pump manufacturer's catalog. The catalog shows the pump curve for each model. The head/flow coordinates can be directly transferred on the manufacturer's graph or the manufacturer's information can be transferred onto a headloss curve. Two points on this graph are important. The segment of the pump curve between the two points it intersects the headloss curves shows the flow that can be expected in the system for the 20 year period for which the system is designed. Figure 5.6 is an example of a manufacturer's pump curve.

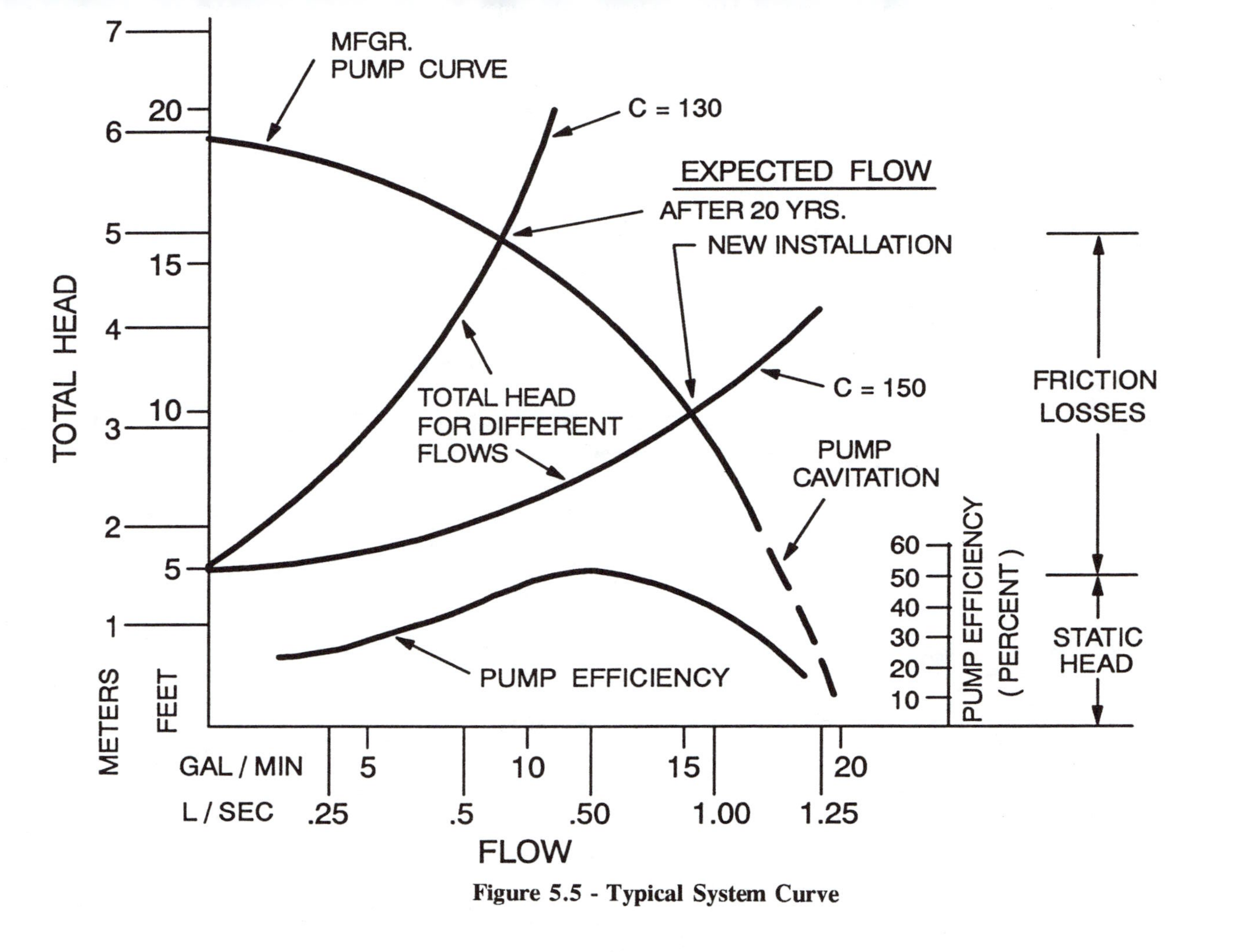

Figure 5.5 - Typical System Curve

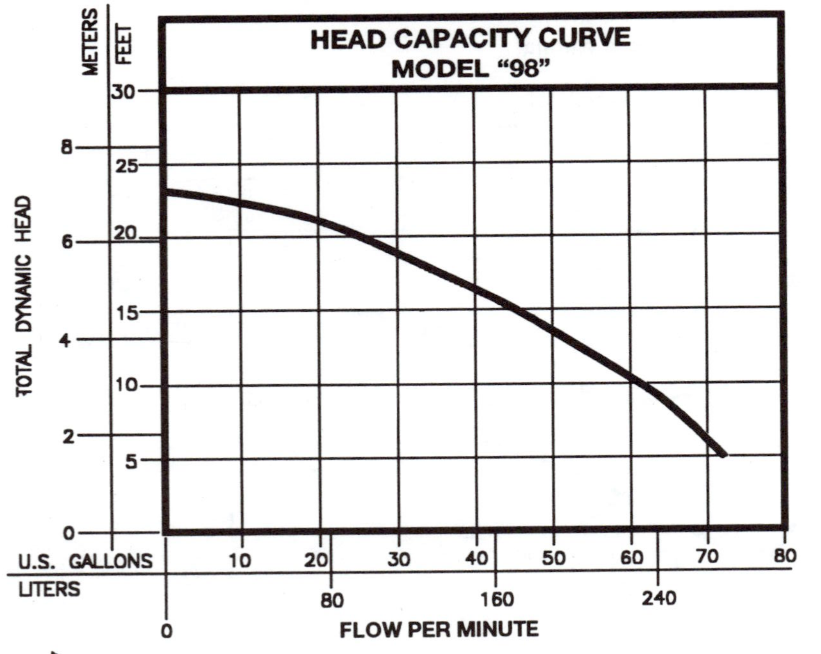

TOTAL DYNAMIC HEAD/FLOW PER MINUTE EFFLUENT AND DEWATERING			
HEAD		CAPACITY UNITS/MIN	
FEET	METERS	GALS	LTRS
5	1.52	72	273
10	3.05	61	231
15	4.57	45	170
20	6.10	25	95
Lock Valve			23′

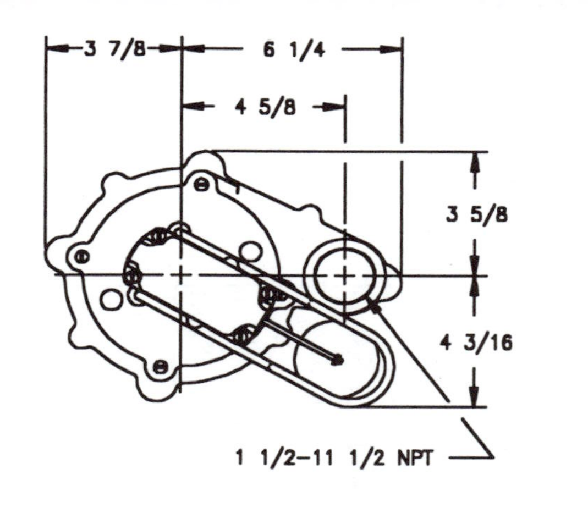

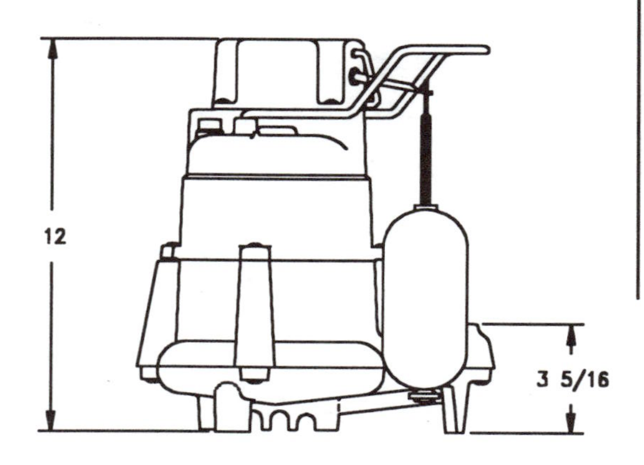

Figure 5.6 - Manufacturer's Pump Curve

DISTRIBUTION SYSTEMS

A pressurized distribution system for an onsite system consists of a pump, a discharge pipe, a distribution manifold (or *header*), and lateral piping. Lateral piping consists of pipes with holes drilled at sizes and locations specified by the designer. The purpose of the system is to distribute the effluent nearly equally among all the perforations in the pipe. Nearly equal distribution will not be obtained until a system curve is developed that accounts for all headlosses. All headlosses cannot be estimated unless every perforation is included in the calculation. The location and size of the holes are based on the principles outlined below.

The basis for equal distribution follows from the concept of *orifice discharge*, which is developed from the principle:

$$C = \frac{ActualFlow}{IdealFlow} = \frac{V}{\sqrt{2g_n H}} \tag{5.12}$$

where
- C = Discharge coefficient (dimensionless)
- V = Actual Velocity, m/sec, (ft/sec)
- H = Total Head at Orifice, m (ft)

Typical values for C are shown in Table 5.4[12].

Table 5.4
Orifice Coefficients

ORIFICE	Sharp Edge	Rounded	Short Tube	Borda
C	0.61	0.98	0.80	0.51

Pressure Distribution

A pressure distribution system is composed of a manifold connected to a series of pipes used to distribute the effluent nearly equally throughout the soil absorption area. There are several controlling considerations in designing a pressure distribution system: hole size, hole spacing and residual pressure. These variables may be set by practice or code, and the proper value for each is critical for the operation of the distribution system.

Hole Sizing. Hole sizing is important for two reasons: small holes can become clogged with grit, debris, and seeds that are not removed in the septic tank. An effluent filtration system can help, but small holes can become plugged. Large holes have high discharge rates; a larger pump will be required to fill the distribution pipe and provide equal flow if larger holes were specified.

Hole Spacing. Hole spacing is also important because holes that are too widely spaced will limit the soil surface area available for effluent treatment. Likewise, too low a residual pressure in the distribution system will result in limited distribution of the effluent in the soil. Generally, holes should not be smaller than 8 mm (0.25 in) unless an effluent filter is used on the septic tank. If a filter is used, 3 mm (0.125 in) holes are acceptable. Newer effluent filters can remove particles as small as 2 mm (0.0625 in) , but holes this small should be avoided unless the quality of the effluent is closely monitored. Holes should be spaced between 0.5 and 1.25 m (1.5 to 2.5 ft) but never more than 3 m (10 ft) apart, the objective being to wet the entire soil surface without overloading any particular location. Excessive spacing will result overloaded soil beneath holes and unwetted areas between holes.

Residual Pressure. The residual pressure should also be a minimum of .75 m (2.5 ft) of head *at the most remote* (or *distal*) *end of the pipe*. This pressure will provide minimal energy to disperse the effluent into the soil. Too high a pressure will create excessive velocity in the effluent through the soil, limiting the time during which to exploit the treatment capabilities of the soil[3].

DERIVATION OF FUNDAMENTAL EQUATIONS

To understand how a pressure distribution system operates, it is essential to review the fundamental equations for flow under a steady condition and use the information in the equations to develop equations specific to orifice flow. Remember that the purpose of pressure distribution is to apply equal volumes of wastewater from each orifice. In practice *equal* flow is impossible because headlosses will increase for each successive orifice. It is possible to provide *nearly equal* flow to each orifice, keeping the differences in flow between the first and last orifices to a specified difference. To see how this is possible, a review of the development of the *Chezy Equation*, the fundamental equation of uniform flow in an open channel, is in order.

Consider a random length, L, of circular open channel. This length of channel is flowing at a constant rate along a uniform slope, S_o. In this channel, which is shown in Figure 5.7, there are four forces identified: F_1 and F_2, which act in equal and opposite directions on each end of the channel length, the weight of the water, W, which acts perpendicular to and parallel to the slope of the channel, the forces exerted by the sides and bottom of the channel, and the frictional stresses,τ_0, that act against the sides and bottom of the channel opposite to the direction of the flow. We are only interested in the forces in the

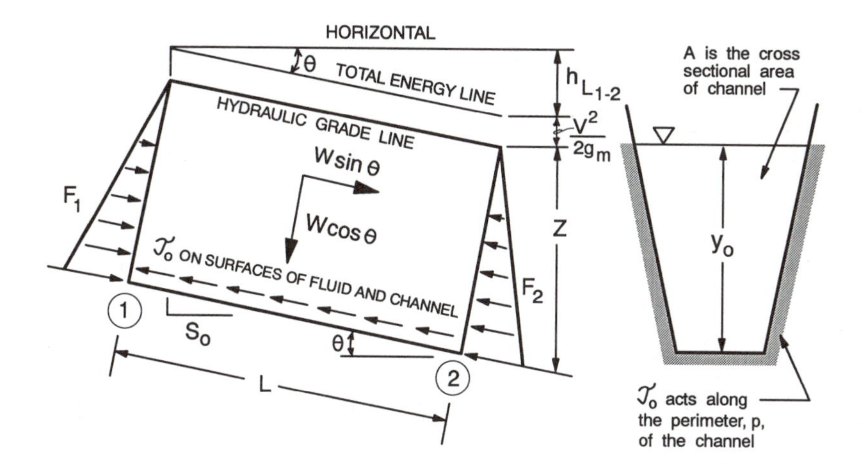

Figure 5.7 - Circular Open Channel

direction of and opposite to the flow. In this case, these forces are F_1 and F_2, $W \sin \theta$, and the frictional forces along the bottom and sides of the channel acting against the flow, $PL\tau_0$. Because the forces are in equilibrium, the sum of the forces equals zero as shown below:

$$F_1 - F_2 + W \sin\theta - P\,L\,\tau_0 = 0 \tag{5.13}$$

In open channels, the slope of the channel, $S_o = \tan \theta$, and for the small slopes encountered in actual channels, $\sin \theta \approx \tan \theta$. If this is the case, Equation 5.13 can be written

$$\frac{\gamma\,A\,L\,h_L}{L} = P\,L\,\tau_o \tag{5.14}$$

where

$h_L/L \;=\; \sin \theta$

$\gamma AL \;=\; W$

Substitution into Equation 5.14:

$$\tau_o = \gamma\,R_h\,S_o \tag{5.15}$$

where

$S_o \;=\; h_L/L$

$R_h \;=\; A/P$

In pipe flow, the frictional stress is related to the friction in the pipe, surface roughness and other values. This relationship can be expressed as follows:

$$\tau_o = \frac{f \rho V^2}{8} \tag{5.16}$$

where

f = frictional factor (a dimensionless constant)

ρ = density of the fluid (kg/m³ or slugs/ft²)

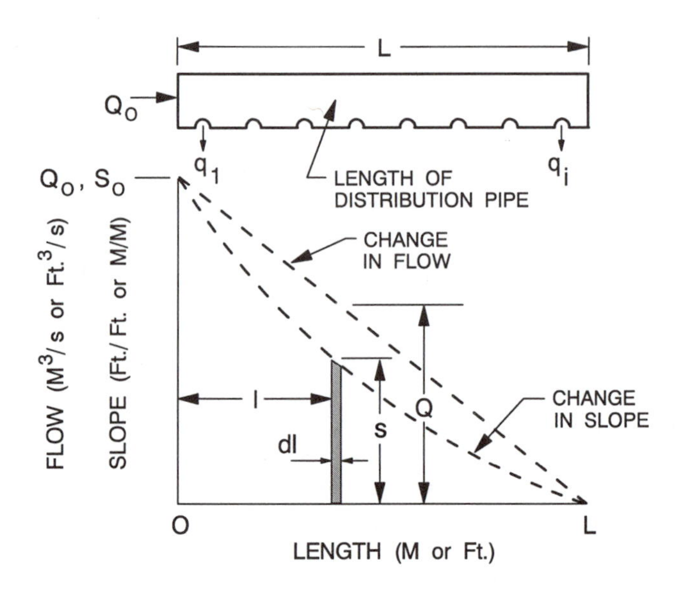

Figure 5.8 - Pressure distribution lateral

However, experience has shown that turbulent flow in regular prismatic channels is similar to flow in pipes; therefore, τ_o for an open channel can equated to τ_o for a pipe. Doing so:

$$\frac{f \rho V^2}{8} = \gamma R_h S_o \tag{5.17}$$

Rearranging and solving for V:

$$V = \sqrt{\frac{8 \gamma}{\rho f}} \sqrt{R_h S_o} = C \sqrt{R_h S_o} \tag{5.18}$$

which is the *Chezy Equation*[14].

In a pressure distribution lateral having perforations of a standard size and separation, the headloss will increase uniformly, at least in an idealized state, as shown in Figure 5.8. Applying the Chezy Equation and keeping in mind that $S_o = h_L/L$:

$$S_o = \frac{V^2}{C^2 R_h} = \frac{Q^2}{A^2 C^2 R_h} = k\,Q^2 \tag{5.19}$$

where

$k = A^2 C^2 R_h$ and is almost constant.

The headloss varies with the length of the pipe and velocity of the wastewater, and thus the flow. For an initial flow of Q_o, the flow at any point l will be in proportion to the length the flow had to travel: $Q_l = Q_o * ((L - l) / L)$. Further, the total headloss is the sum of the headlosses for each element of length:

$$h_L = \int_0^l S\,dl = S\,l \tag{5.20}$$

where l is any point along a length of pipe L.

Because $S = kQ^2$, it can be substituted into Equation 5.19:

$$h_L = \frac{k\,Q_o^2}{L^2} \int_0^l (L - l)^2\,dl \tag{5.21}$$

where k, Q_o, and L are constants

Integrating Equation 5.21 yields

$$h_L = \frac{k\,Q_o}{L^2} \left(L_2 - L\,l^2 + \left(\frac{1}{3}\right) l^3 \right) \tag{5.22}$$

Substituting S_o for Kq_o^2, Equation 5.21 can be written

$$h_l = S_o \left(l - \frac{l^2}{L} + \left(\frac{1}{3}\right) \frac{l^3}{L^2} \right) \tag{5.23}$$

Integrating from *0* to L the total headloss is equal to one-third the headloss for an undrilled pipe:

$$h_L = \left(\frac{1}{3}\right) S_o\,L \tag{5.24}$$

The flow in an orifice can be calculated by rearranging Equation 5.12:

$$Q_{Orafice} = C\,A\,\sqrt{2\,g_n\,H} \qquad (5.25)$$

where
- A = Area, m (ft)
- C = Constant in Table 5.4
- g_n = gravitational acceleration, 9.807 m/sec^2 (32.174 ft/sec^2)

For convenience, and assuming C equals 0.61, Equation 5.25 is rewritten for direct conversion of units for U.S. customary units:

$$Q_{Orafice} = 11.797\,d^2\,H^{0.5} \qquad (5.26)$$

where
- $Q_{orifice}$ = flow, gal/min
- d = orifice diameter, in
- H = total head, ft

The purpose of pressure distribution is to provide *nearly* equal flow through each orifice. The tolerance of the difference in flows should be set by the designer before he begins the design. For a lateral, differences of up to 10 percent are acceptable[15]. The percent difference in flow can be calculated:

The difference in flow between the first and last orifices can be calculated by

$$Percent\ Difference = 100 * \left(\frac{Q_{First\ Orifice} - Q_{Last\ Orifice}}{Q_{First\ Orifice}}\right) \qquad (5.27)$$

where Q can be calculated using Equation 5.25.

For a pressure distribution system to function, the pump switches must be set to provide sufficient volume of wastewater to completely fill the distribution network and supply the dose volume of wastewater. The pump should be able to supply a dose volume that is from five to ten times the volume of the pipe in the distribution system[16,17].

Design Examples

Consider the following design examples using the Bernoulli Equation:

Example 1

Problem:

What is the height of a completely full unpressurized water tower if a pressure gauge at the base of the tower reads 75 lb/in^2?

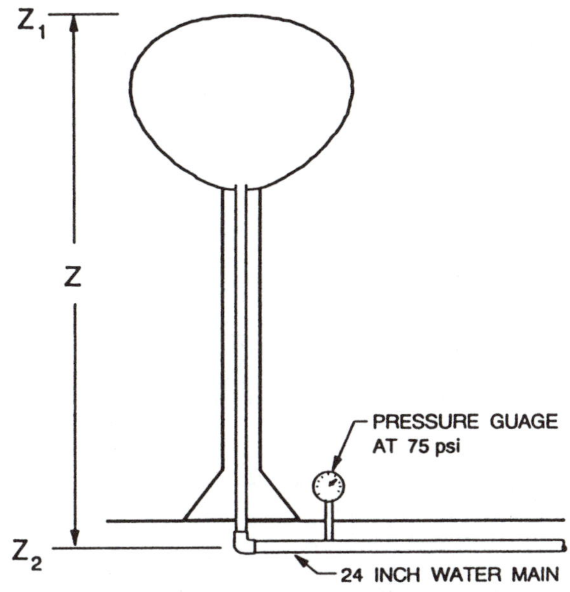

Figure 5.9 - Example 1 Illustration

Solution:

See Figure 5.9. From Equation 5.7, we know that

$P_1 = P_2 = 0$, $V_1 = V_2 = 0$, and $z_2 = 0$

$$\frac{P_1}{\gamma} + \frac{V_1^2}{2g_n} + z_1 = \frac{P_2}{\gamma} + \frac{V_2^2}{2g_n} + z_2 \qquad (5.28)$$

z_1 = (75 lb/in^2) * (144 lb/ft^2 per lb/in^2) * (1/62.34 ft^3/lb)

z_1 = **173.24 ft**

Example 2

Problem:

Suppose that the water tower from the previous problem springs a break in a 24-inch pipe at the base of the water tower. What will the flow out of the pipe be?

Solution:

See Figure 5.10. From Equation 5.7, we know that

$P_1 = P_2 = 0$, $V_1 = 0$, $z_1 = 173.24$ ft, and $z_2 = 0$

173.24 ft = $V^2/2g_n$

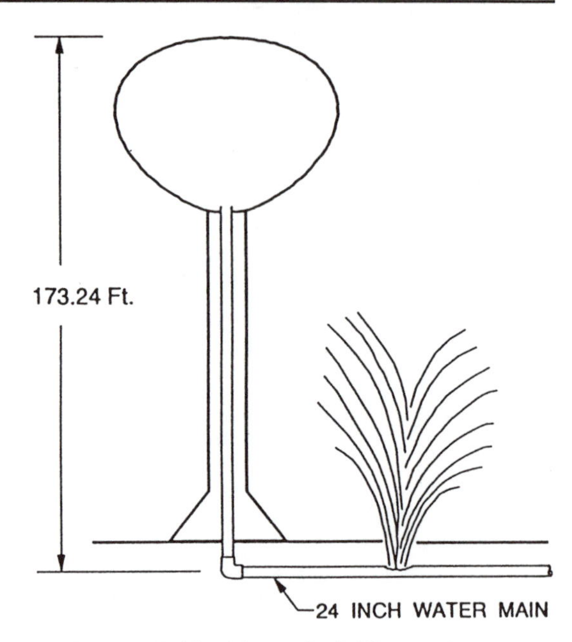

Figure 5.10 - Example 2 Illustration

$V_2^2 = [2 * 32.174 \text{ ft/sec}^2 * 173.24 \text{ ft}]^{.5}$

$V_2^2 = 105.58 \text{ ft/sec}$

$Q = \text{Velocity} * \text{Area}$

$Q = (105.58 \text{ ft/sec}) * [(\pi/4)\ .2^2 \text{ ft}^2]$

$Q = 331.69 \text{ ft}^3\text{/sec}$

$Q = (331.69 \text{ ft}^3\text{/sec}) * (7.48 \text{ gal/ft}^3) * (60 \text{ min/sec})$

Q = 148,862 gal/min (over 214 millions gallons per day!)

Example 3

Problem:

Considering headloss, what is the flow from the break in the 24-inch pipe if the break is treated like a sharp-edged orifice?

Solution:

Orifice flow is the product of the elevation head and the orifice coefficient. We calculated the elevation head and flow in the previous problem. From Table 5.4,

$C = 0.61$

$Q = C * Q$

$Q = 0.61 * 148{,}862$ gal/min

$\mathbf{Q = 90{,}806}$ **gal/min**

Example 4

Problem:

What is the headloss in the 24-inch pipe of the water tower if the main is of worn cast iron pipe and the flow of 90,806 gal/min is confirmed?

Solution:

The headloss in the pipe can be calculated using Equation 5.11. From Table 5.3, C_{hw} = 100, and from the previous problem, Q = 90,806 gal/min. L = 173.24 ft, and d = 24 in.

$$h_L = 0.002083\,L\left(\frac{100}{C_{hw}}\right)^{1.85}\left(\frac{Q_{gpm}^{1.85}}{d^{4.8655}}\right)$$

$h_L = 0.002083 * 173.24 \text{ ft} * (100/130)^{1.85} * (90{,}806^{1.85}/24^{4.8655})$ ft

$h_L = 63.63$ ft

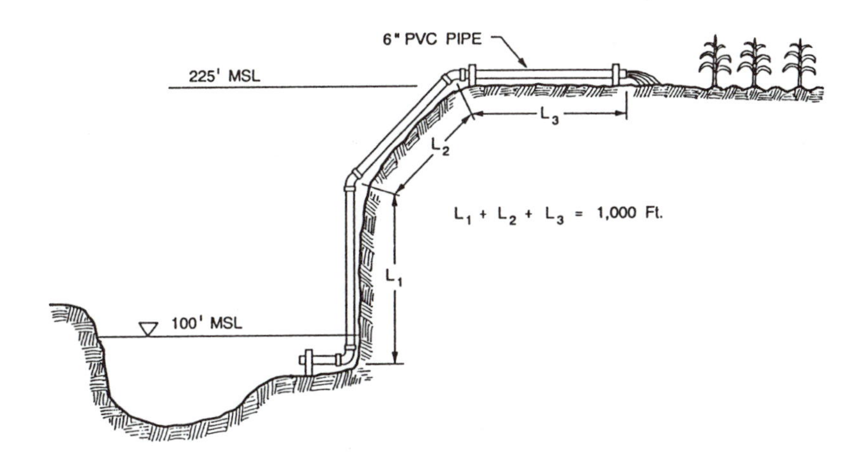

Figure 5.11 - Example 5 Illustration

Example 5

Problem:

It is necessary to pump water through a 4-inch pipe to from a river to a field. The water elevation is 100 ft above sea level, and the discharge elevation is 225 feet above sea elevation. One thousand feet of 6-inch plastic pipe is going to be used. What head must the pump develop to deliver 500 gal/min to the field?

Solution:

See Figure 5.11. Assume that the fitting losses will be negligible. From the information given, Q= 500 gal/min, z_1 = 100 ft, and z_2 = 250 ft. From Table 5.3, C = 150 for new plastic pipe. Using Equation 5.11,

$h_L = 0.002083 * 1{,}000 \text{ ft} * (100/150)^{1.85} * (500^{1.85}/6^{4.8655}) \text{ ft}$

$h_L = 15.8 \text{ ft}$

$V_2^2/2g_n = (500 \text{ gal min}) * (0.134 \text{ ft}^3/\text{gal}) * (.017 \text{ min/sec})$

$\mathbf{V_2^2/2g_n = 1.1 \text{ ft}}$

From Equation 5.6, $P_1 = P_2 = 0$

$V_1^2/2g_n = (15.8 + 1.1 + 250 - 100) \text{ ft}$

$\mathbf{V_1^2/2g_n = 167 \text{ ft}}$

Therefore the designer will have to select a pump that can provide at least 169 ft of head at a 500 gal/min.

Example 6

Problem:

Design a pressure distribution system for a four-bedroom house. The hydraulic loading rate is 0.5 gal/day/ft^2, and the residual pressure at the distal orifice is 2.5 ft. The minimum velocity should be 2 ft/sec, and the maximum velocity should be 6 ft/sec.

Solution:

See Figure 5.12 for a sketch of the system. Establish the flow:

4 bedrooms * 150/gal/day-bedroom = 600 gal/day

Establish the soil absorption area:

600 gal/day ÷ 0.5 gal/day-ft^2 = 1,200 ft^2

Establish total trench length using 3 ft trenches:

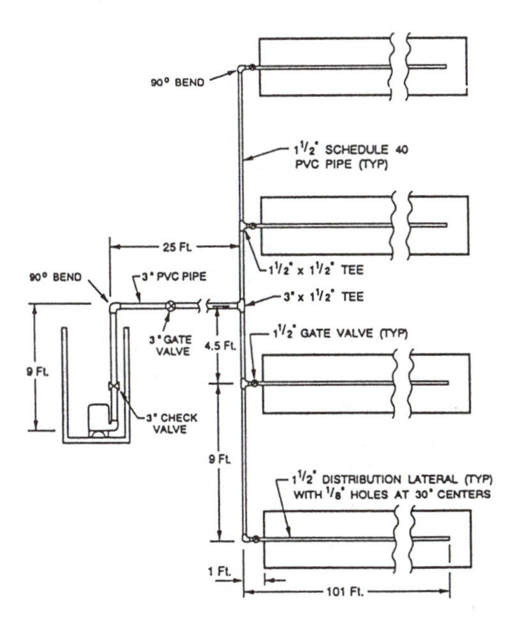

Figure 5.12 - Example 6 Illustration

1,200 ft^2 ÷ 3 ft = 400 ft → **Use four 100-ft trenches**

Calculate the flow from each orifice, assuming a head of 2.5 ft:

$$Q_{orifice} = 11.797 * d^2 * h^{.5}$$

$$Q_{orifice} = 11.797 * (0.125)^2 * (2.5)^{.5}$$

$$Q_{orifice} = 0.291 \text{ gal/min}$$

$$Q_{lateral} = 50 \text{ orifices/lateral} * 0.291 \text{ gal/min-orifice}$$

$$\mathbf{Q_{lateral} = 14.572 \text{ gal/min}}$$

Calculate the wastewater velocity entering the lateral:

$$V = 0.409 * Q \div d^2$$

$$V = 0.409 * (14.572) \div (1.5)^2$$

$$\mathbf{V = 2.649 \text{ ft/sec}}$$

Calculate the headloss per lateral assuming C = 150 for PVC pipe:

$$h_L = \tfrac{1}{3} S_o L$$

$$h_L = \tfrac{1}{3}(.002083 * 100 * (100/150)^{1.85} * [(14.572)^{1.85} / (1.5)^{4.8655}])$$

$$\mathbf{h_L = 0.648 \text{ ft}}$$

Check the difference in flow between the first and last orifice in a lateral:

$$Q_{orifice\ 50} = 11.797 * d^2 * h^{0.5}$$

$$Q_{orifice\ 50} = 11.797 * (0.125)^2 * (2.5)^{0.5}$$

$$Q_{orifice\ 50} = 0.291 \text{ gal/min}$$

$$Q_{orifice\ 1} = 11.797 * (0.125)^2 * (2.5 + 0.648)^{0.5}$$

$$\mathbf{Q_{orifice\ 1} = 0.327 \text{ gal/min}}$$

Table 5.5
Headlosses in Fittings

Fitting	Number	K	Headloss, ft
1½ Gate Valve	4	0.14	0.60
90° Bend	2	0.51	0.11
1½ in Tee	2	0.42	0.09
1½ in Tee	2	1.46	0.55
3 in gate valve	1	0.14	0.02
3 in check valve	1	1.70	7.19
3 in Tee	2	1.26	1.10

Check the percent difference in flow between the first and last orifice:

Percent difference =
[100 *(0.327 — 0.291) ÷ 0.327] = **10 percent**

Calculate the headlosses in the fittings.

Using Equation 5.8 and Table 5.5, the following headlosses are calculated for the fittings. Headlosses in the lateral are calculated using Equation 5.11.

Two 9 ft segments of header pipe will have headlosses of 0.18 ft. Two 4.5 ft segments of header pipe will have headlosses of 0.32 ft. The velocity of the wastewater entering the 1½ in pipe is twice the velocity of a the wastewater entering a lateral; the flow in the pipe is twice the flow in a lateral. The velocity of the lateral was calculated to be 2.64 ft/sec. The maximum velocity in the 1½ pipe will be 5.28 ft/sec.

To keep a lower velocity in the force main to the pressure distribution system, a larger pipe must be selected. A 3 in pipe will have a velocity of 2.64 ft/sec, assuming the flow is to all four laterals.

The final calculations are the headlosses in the 25 ft force main and 9 ft discharge pipe in the pump chamber. Using Equation 5.11, the total headloss is 0.02 ft for the 34 feet of 3 in pipe.

The headlosses for the system are as follows:

Laterals: 4 @ 1.94 ft = 7.76 ft

Fittings: 2.11 ft

1½" Pipe: 2 @ 0.49 ft 0.98 ft

3 in Pipe 0.02 ft

Total headloss: **10.78 ft**

The elevation difference must now be taken into account:

Elevation of pump intake: 91 ft msl

Elevation of laterals: 100 ft msl

Difference: 9 ft

The total head for the system for **C** = 150 is **19.78 ft.**

The calculations should be repeated for C = 140 to estimate the system head after 20 years. There could be as much as a 10 percent degradation in performance. This degradation must be considered during the maintenance cycle.

The dose volume pipe volumes must be calculated and compared:

Dose Volume:

Dose Volume = 600 gal/day ÷ 4 doses/day

Dose Volume = 150 gal/dose

Pipe Volume:

Pipe Volume = $7.48 * (\pi/4) *([127 \text{ ft} * (1\frac{1}{2} / 12)^2) + (34 * (3/12)^2)]$

Pipe Volume = 24.14 gal

Dose Volume/Pipe Volume = 150/24.14

Dose Volume/Pipe Volume = 6.2

Float Switch Setting:

Set Float Switch to pump 150 + 24.14 ≈ 175 gal, assuming 150 gal dose and 25 gal to fill up pipes. The 25 gal pipe volume will drain back into the pump chamber when the pump shuts off.

SUMMARY

A designer must have an understanding of hydraulics in order to design onsite wastewater treatment systems that will function properly. Integral to this understanding is an appreciation of the role of frictional losses that must be accounted for in the selection of pumps, piping, and fittings. As important, the designer must understand how headloss affects the operation of a pressure distribution system. This headloss must be addressed during the selection of orifice size and spacing and in the selection of distribution piping.

To properly select and size pumps, pipes, and fittings, the designer must identify the requirements for the onsite system and the soil and site limitations that affect the design. The designer must then sketch out a preliminary design that is refined as frictional losses are quantified and evaluated against the good practice. If the designer follows these steps,the resulting system should function as intended and provide public health, safety and environmental protection, which are the primary objectives of onsite system design.

References

1. Vennard, John K. and Street, Robert L. *Elementary Fluid Mechanics*, Second Edition. New York: John Wiley & Sons, 1982, pp 1-4.

2. Vennard & Street, pp 281-284.

3. Dunn, I.S., Anderson, L.R., and Kiefer, F.W. *Fundamentals of Geotechnical Analysis*. New York: John Wiley & Sons, 1980, pp 50-56.

4. Vennard & Street, pp 24-26.

5. Dunn, et.al, pp 47-49.

6. Dunn, et.al., p 49.

7. Giles, Ronald V. *Fluid Mechanics & Hydraulics*, Second Edition. New York, McGraw-Hill Book Company, 1962, pp 70-73.

8. Westaway, C. and Loomis, A. (ed.). *Cameron Hydraulic Data*. Woodcliff Lake: Ingersol-Rand, 1984, pp 3-110, 3-115.

9. Westaway & Loomis, p 2-8.

10. Westaway & Loomis, p. 3-8.

11. Westaway & Loomis, p 1-33.

12. Vennard & Street, p 535.

13. Otis, Richard J. *Design of Pressure Distribution Networks for Septic tank-Soil Absorption Systems*, (#9.6). Madison: University of Wisconsin-Small Scale Waste Management Project, 1981, p. 10.

14. Giles, p. 160.

15. Otis, pp 7-8.

16. Metcalf & Eddy. *Wastewater Treatment: Treatment, Disposal, Reuse*, Third Edition. New York: McGraw-Hill, 1991, p. 1067.

17. Fair, Gordon M. and Geyer, John C. *Water Supply and Wastewater Disposal*. New York: John Wiley & Sons, 1954.

CMHC'S HEALTHY HOUSE IN TORONTO

CMHC SCHL
Helping to house Canadians

DRINKABLE WATER AND WASTE WATER MANAGEMENT

DRINKABLE WATER SYSTEM

(P) **Eaves troughs**
Collect roof rainwater, which passes through filter screens and then to cistern.

(R) **Rainwater cistern**
20,000 litres (normally sufficient for 6 months consumption).

(S) **Combination filter**
The rainwater passes through a combination roughing, slow sand, and carbon filter, and then through an ultra violet light disinfection unit before being stored for drinking.

(T) **Drinkable-cold-water tank** [600 L]
Supplies kitchen and bathroom sinks; overflow to reclaimed-cold-water tank.

(O) **Drinkable-hot-water tank** [140 L]
Supplies kitchen and bathroom sinks.

WASTE WATER SYSTEM

(E) **Grey water heat exchanger**

SYSTEM TO RECLAIM WATER

(U) **Septic tank** [3,600 L] (under driveway)
Anaerobic bacteria transforms waste water for treatment by the Waterloo Biofilter.™

(V) **Recirculation tank** [2,000 L] (under floor)
Provides de-nitrification in an aerobic environment.

(W) **Waterloo Biofilter ™**
Aerobic bacteria transforms the effluent to a semi-treated condition.

(X) **Twin combination filters**
Water passes through two combination roughing, slow sand and carbon filters.

RECLAIMED WATER
Water is recycled 3 to 5 times.

(Y) **Reclaimed-cold-water tank** [1,200 L]
Supplies tub, laundry, showers and toilets.

(N) **Reclaimed-hot-water tank** [450 L]
Supplies tub, showers and laundry.

(Z) **GARDEN IRRIGATION**
Site gravel pack disperses overflow water under front garden (about 120 litres per day.)

Legend for Canadian Mortgage and Housing Corporation's "Healthy House in Toronto," courtesy of the Research Division, CMHC. For more information, contact Chris Ives (613-748-2312) or Peter Russell (613-748-2306) CMHC, 700 Montreal Road, Ottawa, ON K1A 0P7 Canada

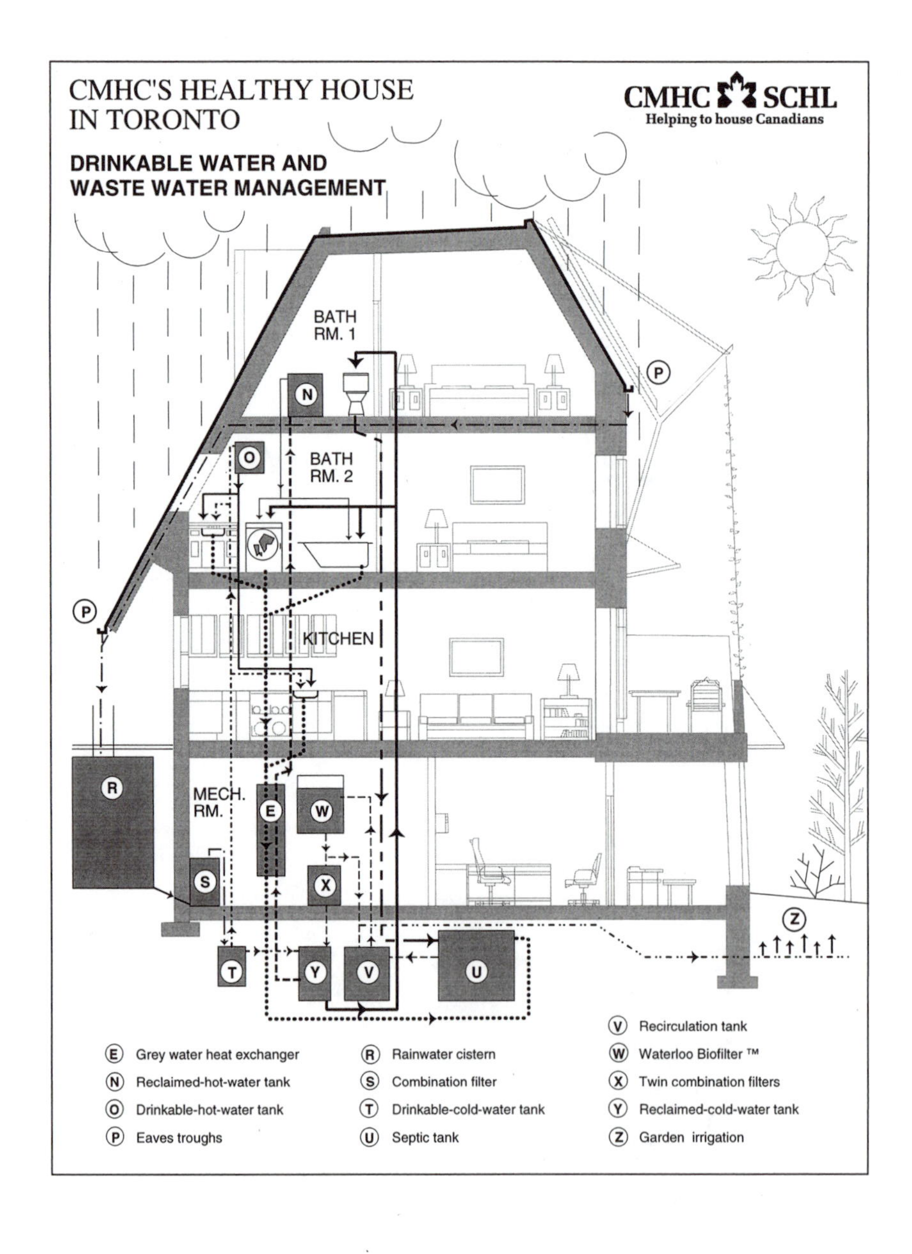

See page 113 for the legend for each water/wastewater component.

CHAPTER 6

WASTEWATER TREATMENT THEORY

Topics of This Chapter

- *Wastewater Process Design Ideas*
- *Typical Wastewater Constituents*
- *Types of Wastewater Treatment Processes*

Wastewater treatment is the process of removing pollutants and pathogens from wastewater, discharging the water to the environment where it is recycled, and disposing of the byproducts of the treatment process. The byproducts include grit, bacterial cells ("sludge"), refuse discarded into the system, gases, and "scum." The ultimate goal of wastewater treatment is to return the water to its original condition than before it was consumed or used. Returning the water to its original condition entails reducing the concentration of organic matter, eliminating the pathogens in the human waste, removing the chemicals discharged into the system, and reducing nutrients to concentrations that do not degrade the environment.

There are five processes in wastewater treatment, all of which are present in practically every wastewater treatment system:

- Mechanical Filtration of Solids,
- Biological Oxidation of Organic Material,
- Disinfection,
- Water Disposal, and
- Byproduct Disposal.

Each serves a specific function and can be accomplished by several different methods. An effective wastewater treatment design is one that achieves the treatment objectives through a balance of the five processes.

MECHANICAL FILTRATION OF SOLIDS

Filtration is the process of separating material by using either a barrier or gravity. The barrier can take many forms, but for most onsite systems, filtration occurs in both septic tanks and in soil absorptions systems. In a septic tank, filtration occurs by two methods: settling and baffles. A septic tank is a "settling basin" - a tank for holding wastewater long enough to allow solids to settle out. When wastewater enters a septic tank, baffles near the inlet pipe of the tank force the wastewater downward. As a result, solids lose their horizontal velocity and drop to the bottom. Oil, grease, and "scum," float to the top because they are lighter than water. The septic tank is designed so the effluent pipe is at the most probable level for drawing off the clarified effluent; solids are trapped below the orifice and scum is trapped above it.

Soil is also a filter. Soil is composed of a three-dimensional matrix of minerals that have pore spaces between them. These spaces are often too small for the passage of solids. In addition, often both the solids and the soil particles have electrical charges. As solids pass through the soil, the solids will either be attracted to or repelled from the soil minerals. As materials adhere to the soil particles, the filtration effectiveness increases. However, if too many fine particles adhere to the matrix materials, the effective filtration may cease and the system may plug.

Soil also serves as a platform for biological growth. Organic material, water, and nutrients support this growth, and this combination of factors has the effect of enhancing the filtration of the wastewater.

MICROBIAL OXIDATION OF ORGANIC MATERIAL

Oxidation Principles

Whenever oxygen combines with another element or compound, the process is called *oxidation*. A common example of oxidation is the rusting of iron:

$$4Fe + 3O_2 \xrightarrow{\text{moisture}} 2\ Fe_2O_3$$

Another example of oxidation can be shown using organic molecules, such as glucose $C_6H_{12}O_6$. Glucose is oxidized in human cells to produce carbon dioxide, water and energy as byproducts:

$$C_6H_{12}O_6 + 6O_2 \longrightarrow 6CO_2 + 6H_2O + \text{energy}$$

However, the modern description of oxidation is more broad; oxidation is the process by which an element or chemical compound *loses electrons*. In the example of the oxidation of iron by oxygen, each iron atom loses three electrons. These electrons go to the oxygen molecule - the *electron receptor*. The concept of an *oxidizing agent* as an electron receptor explains how oxidation of organic substances in an onsite system can occur without oxygen. All that is needed for oxidation is a chemical that will act as an electron receptor. In many anaerobic or anoxic treatment processes, the electron receptors are not oxygen.

Biological oxidation of organic matter can proceed along two possible pathways: aerobic or anaerobic. Aerobic processes use oxygen as their electron receptor and produce biological cells, water, and carbon dioxide as byproducts. Anaerobic processes use nitrate ions and sulfate ions as electron receptor and produce biological cells, carbon dioxide, methane and hydrogen sulfide as byproducts. Each is discussed separately[1].

Aerobic Respiration

Aerobic respiration is the method bacteria use to consume organic material if sufficient oxygen is dissolved in the water. Aerobic treatment consists of two processes; both require oxygen and have new bacterial cells, water, and carbon dioxide as their products. Nutrients, such as compounds of nitrogen and phosphorus, must be available for bacteria to thrive; most domestic onsite systems have sufficient nutrient levels but some special applications may require the addition of nutrients, depending upon the deficiencies. Chapter 3 discusses aerobic processes that occur in nature.

The first step in the aerobic process is the consumption of organic material in the wastewater and is called the *carbonaceous biochemical oxygen demand*. It is generally characterized as:

$$\text{Organic Matter} + O_2 + \text{nutrients} \xrightarrow{\text{microorganisms}} CO_2 + NH_3 + C_5H_7O_2N \text{ (New Cells)}$$

During the process, however, older cells in the reactor are continually dying and their bodies are being used to make new cells. This process, called *endogenous respiration*, is characterized as follows:

$$C_5H_7O_2N + 5O_2 \xrightarrow{\text{microorganisms}} 5CO_2 + 2H_2O + NH_3 + \text{energy}$$

All that remains after this second reaction are the undigestible bacteria hulls, all other products having been assimilated into new bacterial bodies.[2]

In addition, the ammonia that has already been oxidized from organic nitrogen is further oxidized to the nitrate form in a two-step process that involves nitrite as an intermediate product. This process is known as the *nitrogenous biochemical oxygen demand* and is characterized:

$$2NH_3 + 3O_2 \longrightarrow 2NO_2^{-1} + 2H^{+1} + 2H_2O$$

$$2NO_2^{-1} + O_2 \longrightarrow 2NO_3^{-1}$$

Anaerobic Respiration

Lacking oxygen, certain bacteria can use other substances as the electron receptors for energy production: *anaerobic* respiration. In anaerobic respiraton, the major final products are methane and carbon dioxide. However, different bacteria produce different intermediate products. Anaerobic respiration is typically a three-step process.

The first step of anaerobic respiration is *acid fermentation* or *hydrolysis*; it is the decomposition of organic material to complex organic acids. Lipids decompose into fatty acids, proteins into amino acids, and nucleic acids into purines and pyrimidines. Other products include propionates, butyrates, and lactates. Such intensive acid production results in a drop in the pH level and leads to the formation of putrefactive odors.

The second step is *acid regression* or *acidogenesis*. Products of the first step are decomposed into simpler products that will be converted in the third step to methane and carbon dioxide. Collectively, these substrates are known as the *methanogenic substrates*. Byproducts resulting from this step are H_2S, indole, skatole, and mercaptans; all have distinctive odors.

Alkaline fermentation, or *methanogenesis*, is the third step and is the actual release of energy by the conversion of the methanogenic substrates to methane and carbon dioxide.

$$CH_3COOH \xrightarrow{\text{methanogenic bacteria}} CH_4 + CO_2$$

Alkaline fermentation can occur within a pH range from 6.4 to 7.2. Below a pH of 6.2, the methanogenic bacteria are unable to function[3].

WASTEWATER TREATMENT PROCESSES

Wastewater treatment uses bacterial growth to perform five functions:

- remove carbonaceous organic matter,
- nitrification,
- denitrification,
- phosphorous removal,
- conversion of organic wastes to new cell mass, carbon dioxide, and water.

The new cell mass resulting from the process can either be removed from the system and discarded off site or be digested as a part of the treatment process. Water is discharged to the soil or nearby surface water, and carbon dioxide is released to the atmosphere.

There are five types of wastewater treatment processes: aerobic processes, anoxic processes, anaerobic processes, combined processes, and pond processes. These five processes can proceed either as suspended-growth systems, attached-growth, or a combination of the two.

Aerobic Processes

Aerobic processes use aerobic biological processes to achieve treatment objectives. Aerobic treatment processes may be either *suspended growth* or *attached growth* systems. There are essentially four major aerobic processes used for onsite systems: activated sludge, trickling filters, lagoons, and soil. Of these four, activated sludge and tricking filters are used in "packaged" treatment units that may be purchased by manufacturers and lagoons are limited to larger systems serving multiple units. Soil, as an attached-growth system, predominates the design of aerobic treatment units.

Suspended Growth Systems

Activated Sludge. "Activated sludge" refers to any of several types of processes in which bacteria are used — "activated" — to accomplish wastewater treatment via microorganisms. While simple in concept, activated sludge systems require intensive management. Flow into the system is continuous, so either the wastewater must be treated as it is received in a "plug," or newly-received wastewater must be mixed with wastewater that is in the process of being treated. Activated sludge processes can proceed in one of two usual methods. In the Completely Mixed System method, influent wastewater is completely mixed with bacteria and older, partially treated, wastewater in the reactor vessel. In the Plug Flow method, influent is not mixed but flows through the treatment plant as a plug[4].

Completely Mixed Activated Sludge System. Completely mixed activated sludge is based on the idea that treatment will occur if the wastewater is exposed to bacteria in a reactor vessel for a specific length of time. This is the *hydraulic retention time*, θ, which is calculated thus:

$$\theta = \frac{V}{Q} \qquad \textbf{(6.1)}$$

V = Aeration tank volume
Q = Flow into the system

During this time, bacteria will multiply, transforming the organic material in the wastewater into new cell mass, water, and carbon dioxide.

In an activated sludge system, before the wastewater is discharged to the environment, the cells must be removed. The cell mass must be kept in balance, so some cells generated each day must be removed from the system. Therefore, the importance of the hydraulic retention time is replaced by the *mean cell residence time*, θ_c, the average length of time a cell will remain in a vessel. This concept allows the designer and operator a method to achieve process stability by regulating the major factors influencing treatment stability. The mean cell residence time is the ratio of the mass of bacteria in the reactor to the total mass of bacteria removed from the system. For a typical activated sludge design, the mean cell residence time is from five to seven days[5].

The mean cell residence time is calculated thus:

$$\theta_c = \frac{VX}{Q_w X + Q_e X_e} \qquad \textbf{(6.2)}$$

where

θ_c = Average length of time a cell will remain in the reactor vessel (Mean Cell Residence Time)
V = Aeration tank volume (Mgal)
X = Concentration of volatile suspended solids in the reactor vessel, mg/L
Q_w = Flow of Liquid removed ("wasted") from the system
Q_e = Flow from the Clarifier
X_e = Microorganism concentration in the effluent from the clarifier

If the growth and decay rates of the bacteria can be established, the mean cell residence time can be refined, based on these rates, because they reflect both the accumulation and removal of bacteria from the system. These rates can be established empirically by running a series of tests on sample wastewater and measuring the resulting growth and death patterns.

When the operation of an activated sludge system is stable, bacteria will be increasing and dying at different rates, both dependant on the introduction of bacteria and substrate. The relationship between the growth of bacteria and concentration of substrate was explained in Chapter 3 *Microbiolgoy*. The *specific growth rate*, μ, is dependent on the concentration of substrate up to the point where the bacteria are growing at their maximum rate, μ_m.

The point where the bacteria are growing at half their maximum rate, $\mu/2$, is associated with a specific substrate concentration called the *half velocity constant*, K_s[6,7,8]. From this relationship, a ratio between the growth rate and substrate concentration can be shown as follows:

$$\mu = \mu_m \frac{S}{K_s + S} \tag{6.3}$$

where

μ = specific growth rate, 1/time
μ_m = maximum growth rate
Ks = half-velocity constant
S = substrate concentration

The rate of bacterial growth is also dependent on their initial concentration. This relationship can be shown thus:

$$r_{growth} = \mu X \tag{6.4}$$

where

r_{growth} = rate of growth, mass/volume * time
X = concentration of bacteria, mass/volume

Not only do the bacteria grow at rates dependent on the substrate concentration, they also produce different masses of cells depending on the initial concentration of substrate. This makes sense; at lower substrate concentrations, bacteria will need substrate to maintain life, while at higher concentrations, the bacteria will have sufficient substrate to reproduce. The coefficient Y represents the maximum mass of cells formed to the mass of substrate consumed by the bacteria and has the following relationship:

$$Y = \frac{r_{growth}(\textit{maximum mass of cells formed})}{r_{substrate}(\textit{mass of substrate consumed by the bacteria})} \tag{6.5}$$

so, $r_{growth} = Yr_{substrate}$

The ratio between Y and μ_m, which is the maximum rate of bacterial growth, is replaced by a constant, k, and the relationship among the three is:

$$k = \frac{\mu_m}{Y} \tag{6.6}$$

Based on this value, the rate of substrate utilization can be written thus:

$$r_{su} = \frac{\mu_m X S}{Y(K_s + S)} = \frac{kXS}{K_s + S} \tag{6.7}$$

When bacteria death is subtracted from this value, the mean cell residence can be calculated:

$$\theta_c = \frac{1}{Y\left(\frac{kS}{K_s + S} - k_d\right)} \tag{6.8}$$

where

θ_c = mean cell residence time, days
Y = yield coefficient (mg of volatile suspended solids/mg BOD_5)
k_d = endogenous decay coefficient, 1/days
S = substrate, mg/L

Extended Aeration

Bacterial growth goes through successive phases that have been well documented and characterized. The terminal phase in bacterial growth is the *endogenous phase* when the food supply is insufficient to sustain the colony and the bacteria begin to die. During this phase, surviving bacteria can consume the nutrients released by the dead bacteria. This process can be exploited by the designer by reducing the *loading rate,* that is, the concentration of substrate, to the system and prolonging the mean cell residence, θ_c, time to approximately 10 days[9].

Plug Flow Activated Sludge System

A plug flow activated sludge system proceeds by introducing wastewater into a reactor where it is held for a specific period, based on the design of the individual reactor. During this period, the plug of wastewater is treated, then discharged to a clarifier where the bacteria settle out and the clarified wastewater or *effluent* is then discharged. The settled-out and collected bacteria can either be recycled into the reactor or discharged to a solids-handling unit for disposal. A common approach is to recycle some bacteria back to the reactor to enhance treatment.

Plug flow systems are efficient, but because of the problems of irregular flow, shock loads of toxic substances, and inherent difficulties of maintaining process stability, completely-mixed systems are often favored in design.

A completely mixed activated sludge system differs from a plug flow system because there is no mixing of partially-treated and untreated wastewater in a plug flow system. The lack of mixing, as noted before, makes a plug flow system more efficient, but in practice, a difficult condition to maintain. The lack of mixing makes plug flow systems more suspectable to impacts from shock loads.[10]

$$\frac{1}{\theta_c} = \frac{Yk(S_o - S)}{(S_o - S) + (1+\alpha)K_s \ln\frac{S_i}{S}} - k_d \qquad (6.9)$$

$$S_i = \frac{S_o + \alpha S}{1 + \alpha} \qquad (6.10)$$

where

S_i = Influent BOD Concentration
S = Effluent BOD Concentration
S_o = Influent BOD Concentration after Dilution
α = Recycle Ratio

Sequencing Batch Reactor

A sequencing batch reactor is a type of suspended growth activated sludge system. What distinguishes a sequencing batch reactor from other systems is that all of the reactions occur within a single tank. Traditional activated sludge systems have separate vessels for treatment and settling. All sequencing batch reactors use the same five steps: fill, aerate, settle, decant, and idle in the same reactor vessel[11].

Aerated Stabilization Lagoon

An aerated stabilization lagoon is really a type of extended aeration activated sludge design, except that the basin is usually an earthen berm, and oxygen is supplied by diffusion or blowers. In an aerobic lagoon, bacteria are kept in suspension by use of mixers[12]. The removal of soluble BOD can be estimated using the following equation:

$$\frac{S}{S_o} = \frac{1}{1+\left(\frac{kX}{(K_s+S)}\right)\left(\frac{V}{Q}\right)} \tag{6.11}$$

Attached Growth Systems

Attached growth systems operate by passing wastewater over and through a medium on which the bacteria can attach themselves and multiply. The medium can be either natural, porous stones, or synthetic, such as slag or plastic rings. The media can be stacked in a tower or placed in a concrete tank.

Because there is no need to either continuously monitor the concentration of bacteria or maintain suspension, attached growth systems are inherently easier to operate. For this reason, attached growth systems, chiefly sand filters, are often preferred to suspended growth systems for individual, cluster, and small community systems.

Tricking Filter

A trickling filter consists of a bed of porous material upon which organisms attach themselves and grow, developing a *slime layer*, through which the wastewater percolates. Trickling filter media can be rocks, slag, plastic media, or any other medium that presents a high surface area. The trickling filter has an underdrain that is used to collect the percolate and any solids that may have washed through the system. The underdrain also provides a conduit for oxygen transfer throughout the system.

Most of the treated water is discharged after it has been clarified. A portion is recycled back through the system where it serves two purposes: diluting the influent wastewater and keeping the biological slime layer moist.

Trickling filters can be designed according to theoretical or empirical equations. The theoretical equation is a mass balance of organic material across a slime layer.[13,14]

The resulting equation is:

$$r_s = \frac{Ehk_o\bar{S}}{K_m + \bar{S}} \tag{6.12}$$

where

r_s = rate of flux of organic material into the slime layer,
E = effectiveness factor $(0 \leq E \leq 1)$
h = thickness of slime layer, meters
k_o = maximum reaction rate per day
$\bar{S}$ = average BOD concentration in the bulk liquid in volume element
K_m = half-velocity constant

The recirculation factor, a measure of how many times the wastewater passes through the medium, can be calculated as follows:

$$F = \frac{1+R}{\left(1+\frac{R}{10}\right)^2} \tag{6.13}$$

where

R = Recirculation Ratio (Q_r/Q)
Q_r = Recirculation Flow
Q = Wastewater Flow

Trickling filters based on theoretical designs have not worked well. Instead, many designers, in an attempt to adequately size tricking filter beds, use equations based on performance of working filters. An equation developed by the National Research Council (NRC) is commonly, and dependably, used[15]. The NRC formula is:

$$E_1 = \frac{100}{1+0.0561\sqrt{\frac{W}{VF}}} \tag{6.14}$$

where

E_1 = Percent efficiency of BOD removal at 20°C, including recirculation
W = BOD Loading, lb/day
V = Volume of Filter Media, 1000 ft^3
F = Recirculation Factor

Rotating Biological Contactors

A rotating biological contactor (RBC) is a series of closely-spaced circular plastic disks. The disks are submerged and rotated through wastewater and then exposed to the atmosphere. This rotation allows the microorganisms to alternately be exposed to the wastewater and oxygen so that organic material can be assimilated by the aerobic bacteria. Microorganisms become attached to the disks, specially designed to support the growth, and eventually form a slime layer.

RBC units are designed based on pilot studies, but the designs can also be based on empirical formulas, much like trickling filters. The current practice is to design units based on the pounds of soluble BOD per thousand square feet of RBC surface area. If nitrification is expected, the pounds of ammonia per thousand square feet of surface must also be factored into the design.[16]

Sand Filters

Sand filter designs are based on four important investigations, conducted between 1887 and 1908 at the Lawrence Experimental Station. The current design practice is based on those observations made from empirical relationships developed from these studies. These observations and relationships were refined in three subsequent studies[17].

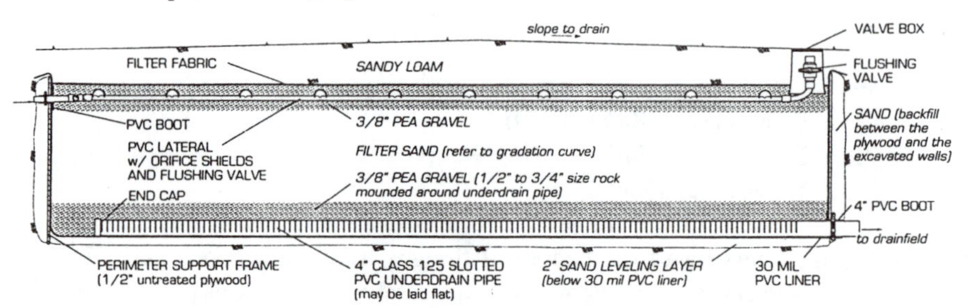

Figure 6.1 - Sand filter

Sand filters, similar to the one shown in Figure 6.1, achieve wastewater treatment in a biological mat that develops below the surface of the filter. Most of the treatment occurs within the first nine to 18 inches of filter depth, although more consistent performance is achieved by using deeper filters. Sand filters are most effective in removing BOD. They perform some nitrification, but effective nitrification can be achieved only by either recirculating the effluent or using a second filter. Current design practices include loading rates between 1.0 and 6.0 gallons per day per square foot of area (gpd/ft^2) and depths between 24 and 42 inches. Sand size and uniformity affect the loading rate. Maintenance requirements have a bearing on sand filter design. The biological mat will require periodic raking, the frequency is a function of the hydraulic loading and the particle sizes of the sand. As the hydraulic loading increases, there will be a need for more frequent raking.[18]

Anoxic processes

Anoxic processes are aerobic respiration processes in which free oxygen is not available or where the bacteria are stressed to consume more nutrient than normal. Nitrogen removal is accomplished by the first process, and phosphorous removal is accomplished by the second.

Nitrogen Removal

The reduction of nitrate to nitrogen gas, *denitrification*, can be accomplished by facultative heterotrophic bacteria that reduce nitrate to nitrogen gas in a two-step process. The first step is the conversion of nitrate to nitrite; the second step is the conversion of nitrite to nitrogen gas, the oxygen already stripped away for aerobic respiration. Other products of the second reaction can be nitrous oxide and nitric oxide.

Denitrification is sensitive to both oxygen availability and pH. The process will be suppressed if oxygen is dissolved in the wastewater and will be reduced if the pH is above 8 or less than 7. The process will not occur if there is no carbon source. If the carbon is not available, it must be introduced into the system either by using methanol, acetate, or by recycling treated wastewater with untreated wastewater.[19]

Phosphorous Removal

Phosphorous and oxygen atoms can exist in a number of different numerical and geometric configurations. Typically, phosphorous is in wastewater as orthophosphate, polyphosphate, and organic phosphorous. In the activated sludge process, approximately 30 percent of the influent phosphorous can be removed by bacteria that use this phosphorous for cell synthesis. Certain bacteria exhibit an ability to ingest additional phosphorous if they are put through an alternating sequence of aerobic and anoxic conditions. The phenomenon, known as luxury uptake, can be used to remove excess phosphorous from the wastewater[20].

Anaerobic processes

Apart from their application as septic tanks, anaerobic processes are typically used to treat sludges and other high-strength wastes. The anaerobic process is described in Chapter 3 and involves two basic processes - liquefaction and gasification. Liquefaction occurs with extracellular enzymes that hydrolyze complex carbohydrates to simple sugars, proteins to peptides and amino acids and fats to glycerol and fatty acids. The ultimate products of the liquefaction process are primarily volatile organic acids produced by "acid producing" strains of bacteria.

Anaerobic digestion involves three stages: acid fermentation, acid regression and alkaline fermentation. During the acid fermentation stage, carbohydrates are broken down to low molecular weight fatty acids - primarily acetic, butyric and propionic. A drop in pH results from the acid formation, and those products having offensive odors are released to the atmosphere. Only about 10 percent of the grease present is degraded during this stage. During the acid regression stage, decomposition of organic acids and soluble nitrogenous compounds occurs, resulting in the formation of ammonia, amines, acid carbonates, CO_2, N_2, CH_4, and H_2. The pH rises and H_2S, indole, skatole and mercaptans are produced. During the alkaline fermentation stage, complete destruction of cellulose and nitrogenous compounds occur. Organic acids produced in the early stages are then broken down to CO_2 and CH_4. The organisms primarily responsible for the process are the spore-forming anaerobes, methane bacteria, and fat-splitting organisms.

Many anaerobic designs, whether attached growth or suspended growth, use an upflow design; that is, the wastewater flows into the bottom of the reactor, travels upwards and is discharged at the top. An upflow pattern ensures that the media will continually be submerged in an anaerobic condition and that maximum bacteria/wastewater contact is achieved[21].

Upflow Anaerobic Sludge Blanket

An upflow anaerobic *sludge blanket*, as shown in Figure 6.2, is an example of a suspended growth unit. The process operates by having the wastewater pass from the bottom of the reactor through a vessel filled with biologically-formed granules. As the wastewater passes in and around the granules, the anaerobic bacteria consume the waste and generate byproducts. Because the granules must be kept in suspension by the action of the wastewater, the flow into the reactor must be kept high.[22]

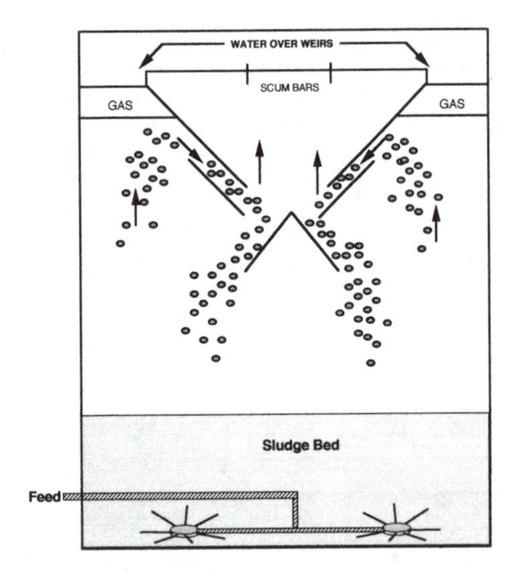

Figure 6.2 - Upflow anaerobic sludge blanket
(Courtesy of Biothane®, Camden, NJ)

Upflow anaerobic filter

An *upflow anaerobic filter* operates the same way as a sludge blanket, with two notable exceptions. The anaerobic filter uses media to support the biological growth. Consequently, the flow into the filter can be lower than into a sludge blanket[23].

Expanded bed

An *expanded bed* treatment unit is composed of elements of both attached and suspended growth processes. In this process, wastewater passes upward through media that is "expanded" (raised) by hydraulic pressure generated by the flow. The biomass grows on the expanded media[24].

Ponds

Ponds are large basins for holding and treating wastewater. They can be classified according to the presence or absence of oxygen. Aerobic ponds are areas surrounded by a berm; this enclosed area is used to treat wastewater through natural processes involving algae, bacteria or both. If the objective is to stimulate algae growth, the ponds are kept shallow, perhaps no deeper than 1.5 feet. If the objective is to maximize oxygen production, the pond depth may be increased to 5 feet. If the deeper pond is used, the best treatment occurs when the pond is periodically mixed with a blower, an aerator, or pumps.[25]

Aerobic ponds treat wastewater effectively, but the tradeoff is that the effluent will be loaded with algae and bacteria that will create a higher BOD; therefore, the effluent must be clarified before it is released. Clarifying lagoon effluent can be difficult, since algae can quickly overload simple filters and do not easily settle out.

Facultative ponds use a combination of aerobic and anaerobic processes to treat the wastewater. A facultative pond is a large area surrounded by an earthen berm that receives raw sewage and forms three zones: aerobic, anaerobic, and facultative. When the wastewater enters the pond, solids will settle to the bottom where they form an anaerobic zone. The surface will remain aerobic due to the algae and re-aeration, while the intermediate facultative zone will form between the aerobic and anaerobic zones.[26,27]

RUCK System

Standard septic systems remove about 10% of the nitrogen from wastewater. The RUCK system provides for some additional treatment of wastewater, passively removing nitrogen and phosphorus. For this system, the two types of wastewater — blackwater (toilets) and graywater (kitchen and laundry) — are separated. Table 6.1 shows the difference between these two types of wastewater. The blackwater, rich in nitrogen compounds, flows into a standard septic tank, the effluent going through a three-stage aerobic sand filter.

Table 6.1
Blackwater and Greywater Constituents

Constituent	Blackwater	Greywater
COD	135	75
BOD_5	40	45
SS	40	25
TN	12	1
PO_4^{-3}	4	4
Chlorides	16	-
Grease	-	15
Temp °C	20	50
Flow %	40	60

The greywater is treated in a different septic tank, then the effluents from both systems mix in a fixed-medium anaerobic reactor. This process results in denitrification. Figure 6.3 shows the design of the RUCK system. The system removes 80 - 90% of the nitrogen from the total wastewater before it goes to the soil for infiltration.

Usually, the plant nutrient phosphorus is immobilized in the soil infiltration system. However, in geographical areas where the immobilizing minerals do not naturally occur and where waterbody protection from excess phosphorus is important, the RUCK system-and most other standard systems- can be adapted to immobilize the phosphorus[28].

Phosphorus removal has been achieved by either adding a phosphorus precipitant via a toilet flush or by loading the final filtration medium with bauxite or another substance that combines with and immobilizes soluble phosphorus.

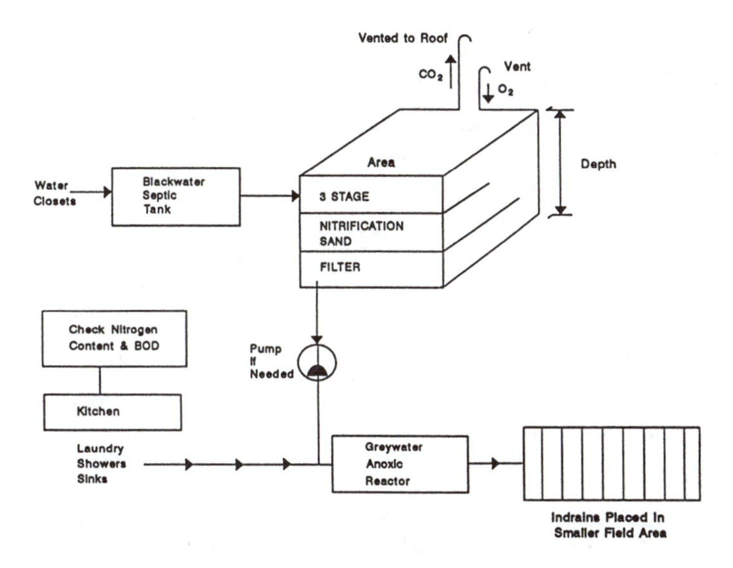

Figure 6.3 - The RUCK system

References

1. Gaudy, A. and Gaudy, E. *Microbiology for Environmental Scientists and Engineers*. New York: McGraw-Hill Book Co., 1980, pp 418, 519.1.5

2. Gaudy and Gaudy, pp 50-72.

3. Gould, R. (ed.). Anaerobic biological treatment processes. *Advances in Chemistry Series 105*. Washington D.C.: American Chemical Society.

4. Metcalf & Eddy, Inc. *Wastewater Engineering Treatment, Disposal, and Reuse*. New York: McGraw-Hill, 1991, pp 378-403.

5. Qasim, S. *Wastewater Treatment Plants Planning, Design, and Operation*. New York: Holt, Rinehart and Winston, 1985, pp 303-326.

6. Lehninger, A. *Principles of Biochemistry*. New York: Worth, 1982.

7. Mazur, A. and harrow, B. *Textbook of Biochemistry*, 10th Ed. Philadelphia: W.B. Saunders, 1971.

8. Gaudy and Gaudy, pp 232-303.

9. Metcalf and Eddy, pp 529-553.

10. Parker, D. "Wastewater Process Design." University of Arkansas, Fall 1985.

11. U.S. Environmental Protection Agency. *Sequencing Batch Reactors*, EPA/625/8-86/011. Cincinatti: Office of Technology Transfer, 1986.

12. Metcalf and Eddy, p 605.

13. Atkinson, B, Davies, I, and How, S. "The Overall Rate of Substrate Uptake by Microbial Films," parts I and II, *Trans. Inst, Chem. Eng*, 1974.

14. Metcalf and Eddy, pp 407-408.

15. National Research Council. "Sewage Treatment at Military Installations," Report of the Subcommittee on Sewage Treatment in Military Installations, National Research Council. *Sewage Works J.*, Vol 18, No. 5, 1946, pp 787-1025.

16. U.S. Environmental Protection Agency. *Process Design Manual for Upgrading Existing Wastewater Treatment Plants*, EPA-625/1-83/015. Washington D.C.: Office of Technology Transfer, 1974.

17. Dunbar, Dr. *Principles of Sewage Treatment* Charles Griffin & Company, Limited, London, 1908. p.121

18. Sauer, D. *Intermittent Sand Filtration of Septic Tank and Aerobic Unit Effluent Under Field Conditions*. Master's Thesis. University of Wisconsin-Madison, 1975.

19. Gaudy and Gaudy, pp 336-339.

20. U.S. Environmental Protection Agency. *Phosphorous Removal*, EPA/6251-87/001. Cincinatti: Center for Environmental Research Information, 1987, pp 15-18.

21. Barker, H. *Bacterial Fermentations*. New York: John Wiley and Sons, 1956.

22. Metcalf and Eddy, pp 427-428.

23. Mitchell, D. *Laboratory and Prototype Onsite Denitrification by an Anaerobic-Aerobic Fixed Film System*. Fayetteville: University of Arkansas, Department of Civil Engineering, 1989.

24. Metcalf and Eddy, p 429.

25. U.S. Environmental Protection Agency. *Design Manual, Municipal Wastewater Stabilization Ponds*, 1983.

26. Thirumurthi, D. "Design of Waste Stabilization Ponds." *J. San. Eng. Div.*, ASCE, Vol. 95, No. SA2, 1969.

27. U.S. Environmental Protection Agency. *Upgrading Lagoons*. Washington D.C.: Office of Technology Transfer, 1973.

28. Laak, Rein. *Wastewater Engineering Design for Unsewered Areas*. Technomic Press, Lancaster, Pa. 1986.

CHAPTER 7

ONSITE WASTEWATER TREATMENT ALTERNATIVES

Topics of This Chapter

- *Onsite Wastewater Treatment System Operations*
- *Traditional Design Alternatives*
- *Innovative System Designs*

Over the last century, many types of onsite wastewater treatment systems have been designed and used. Because of the enormous variations in temperature, soils, water use, availability of electrical power, and climate throughout the world, no one design has been used everywhere. Instead, different treatment and disposal alternatives can be combined to produce optimum solutions for particular sites. Onsite wastewater treatment systems range from the traditional septic tank/soil absorption system to designs using features such as pressure distribution of effluent, sand filters, and aeration. New treatment technologies are being developed, but the basic principles of wastewater treatment remain constant. The complexity of the system required for a particular site varies with the infiltrative capacity of the soil, geology, climate, topography of the site, and regional code requirements. For example, in a temperate climate and loamy soil, a "traditional" septic system may operate well for generations. However, in a cold climate on a sloping site with clayey soils, a more complex system may be needed — perhaps pressure distribution and fill.

TRADITIONAL SYSTEMS

In most localities, all onsite system designs are judged in comparison with the basic septic tank/soil absorption system. So prevalent is this view that the septic tank/soil absorption system is the "conventional" system, from which all other designs are known as "alternative" designs. Because of its significance, the septic tank/soil absorption system must be fully understood.

Septic Tank

A septic tank is a buried, watertight container used to clarify and partially treat wastewater. The septic tank has been in used in one form or another for over 100 years. The septic tank was originally designed to serve as a settling basin—

a vessel to separate scum and grit from the liquid — the effluent from the tank then sent to a sewer or soil for disposal. The clarification function of the tank was known, but the biological processes that partially digested the sewage were discovered almost by accident. Scientists found that the organic solids in the wastewater decomposed if they stayed in the tank long enough[1].

The septic tank is designed to accomplish two major tasks — clarification and treatment. Clarification is a function of the detention time and water extraction method. Solids settle out of the water based on size and specific gravity; smaller, lighter particles take longer to settle than heavier particles. Clarification also includes the removal of fats, oils, and greases, which float to the surface along with soap suds and "scum." The variable features of a septic tank are: size, shape, number of chambers, number and style of baffles and gas venting provisions. Figures 7.1 show a traditional septic tank in cross section.

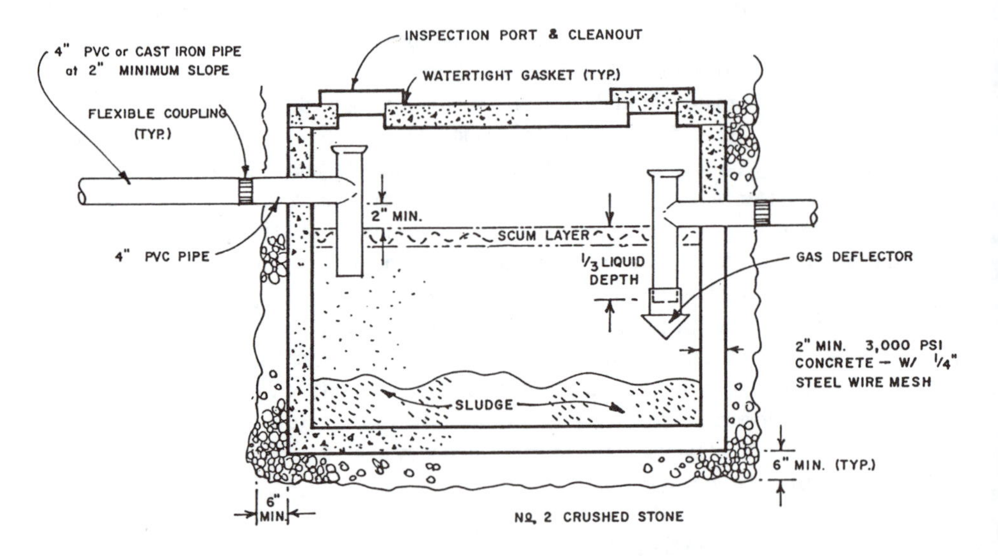

Figure 7.1 - Single compartment septic tank

Treatment consists of two components: clarification by physical treatment and biological treatment by anaerobic digestion. As described in Chapter 3, *Microbiology of Wastewater and Soil*, anaerobic treatment partially decomposes the organic matter into simpler compounds that can be treated further in the septic tank or discharged into the soil for aerobic treatment.

Tank Sizing. Tank size and household water usage determine the detention time — an important principle of septic tank size selection. Efficient clarification takes time to complete because fats, oils, greases, and suspended solids travel slowly in water and may require hours to either float to the top or settle to the bottom.

A septic tank also accomplishes treatment through the biological activity of anaerobic or facultative bacteria. This type of biodegradation takes many hours to work, so treatment efficiency is linked to detention time. The recommended detention time ranges from 36 to 48 hours, but the absolute minimum is 24 hours[2,3]. To determine the tank size, divide the desired detention time by 24 hours and multiply that number by the amount of wastewater generated per day:

$$\text{Tank Size} = \left(\frac{\text{Detention Time (hours)}}{\text{24 hours/day}}\right) * \text{Volume of Wastewater per Day} \qquad (7.1)$$

For example, many regulatory codes specify a minimum septic tank capacity of 2,800 L (750 gal) for a three-bedroom house. These same rules specify a design flow of 1,700 L/day (450 gal/day) for a three-bedroom house, 560 L (150 gal) per day per bedroom (assuming there are two occupants per bedroom, each using 280 L/day (75 gal/day). The theoretical detention time based on the septic tank capacity divided by the design flow is 1.67 days or 40 hours, which is within the design limits. In reality, the average 3-bedroom house has three people, actually using about 185 L/day (50 gal/day), so the assumptions for design are very conservative, so more than the minimum treatment usually occurs.

A septic tank must serve as a receptacle for all the settleables and floatables for two to four years of accumulation. For this reason, the septic tank design must include provisions for adequate storage. The storage capacity is based on the intended use of the tank and anticipated pumping interval. A tank that is too full of solids will not function properly and will allow unwanted substances to pass straight through to the soil absorption system and damage it.

The well-designed tank can provide for removals of total suspended solids (TSS) from 60 percent to 90 percent and biochemical oxygen demand (BOD) removal from 30 percent to 80 percent[5,6]. The septic tank will continue to perform its duties until the limits of its storage capacity are reached. Then, the solids will have to be pumped out of the tank, the frequency of which is determined by the number of people using the system and the use of the system. Table 7.1 shows typical recommended pumping frequencies.

Tank Configuration. The shape of the tank must be designed to maximize the detention time of the wastewater. Surface area is more critical for settleability than depth, so a shallow, wide tank is preferable to a deep, narrow tank if both have the same capacity. Septic tanks tend to be shallow, six feet a common depth, because the tank is easier to transport and install and poses less of a saf-

Table 7.1
Estimated Septic Tank Inspection and Pumping Frequency in Years[4]

	Number of People Using the System				
Tank Size Liters (Gallons)	**1**	**2**	**4**	**6**	**8**
3,400 (900)	11	5	2	1	<1
3,785 (1,000)	12	6	3	2	1
4,542 (1,200)	16	8	3	2	1
5,700 (1,500)	19	9	4	3	2

ety risk because the content level of the tank is not much deeper than the height of an average person.

An improperly configured tank will allow wastewater to "short-circuit" through the tank to the outlet. An example of this phenomenon occurs in restaurants where hot water is used. Hot water has a lower density than the cooler water that has been in the septic tank longer. Because of this density difference, the warmer water will float on top of the cooler water. If the inlet and outlet are in a straight line, it is possible for the warm water to travel straight through the tank with little or no treatment. Short circuiting can also allow solids to migrate to the absorption field if the wastewater is not given sufficient time for the solids to settle out. Baffles and compartments eliminate the possibility of this straight-through path.

Baffles. Baffles are critical to the success of the septic tank and the septic tank will not function without them. Influent baffles restrict and redirect the flow of the incoming wastewater to prevent short-circuiting. By doing so, baffles control the flow of the settleable and floatable materials. Effluent baffles prevent floatables, scum, or suspended solids from flowing into the drainfield. Baffles come in many sizes and styles, the simplest just a bend and extension in a pipe. Baffles can also be concrete or fiberglass partitions attached to the ceiling of the tank. No significant difference has been found between using plastic and concrete baffles[7]. More elaborate baffling systems can include pipes, screens and weirs that direct and/or filter the flow.

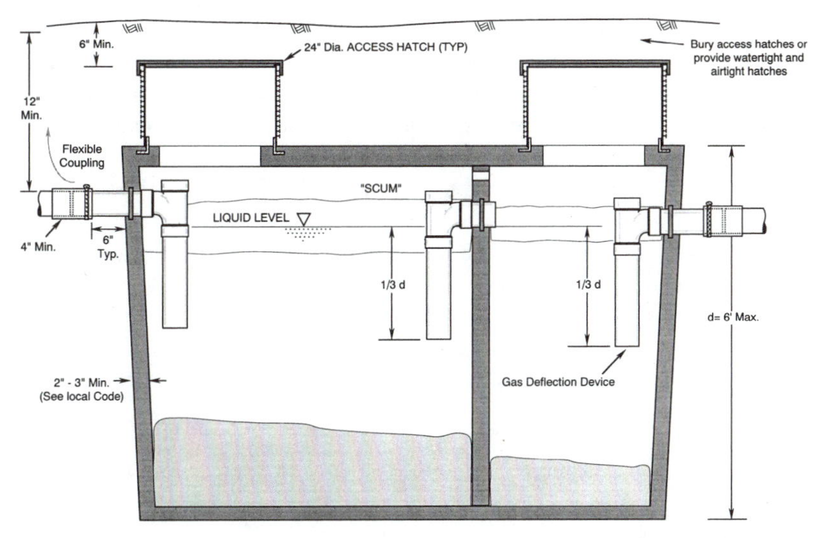

Figure 7.2
Compartmentalized Septic Tank

Gas Management. Gases that are produced in the septic tank must be vented out of the system. Anaerobic processes result in the formation of hydrogen sulfide (H_2S), methane (CH_4), and other putrescent gases. The first step in gas management is to seal the septic tank against escaping gases by burying the access hatches or providing them with rubber gaskets. An open top to the inlet and outlet pipes provides for ventilation of the tank to the vent stack of the structure and the soil absorption system. Through these outlets, gases can circulate and escape from the tank. Still, gases in a septic tank are present in lethal — and sometimes explosive — concentrations, and no one should enter an operating septic tank without supplementary breathing provisions.

Sometimes, gaseous products of anaerobic digestion are generated in the solids at the bottom of the septic tank and large bubbles can belch up from the solid layer. When belching occurs, the gas disrupts the solid layer and can allow some solids to migrate to the outlet pipe and be carried to the drainfield. If this happens often, the soil absorption area can become clogged with solids. To prevent this occurrence, the outlet pipe can be fitted with a deflector to prevent the gases from travelling up the pipe. The deflector can be a 45° bend or a conical device manufactured to fit over the outlet pipe. The use of other baffling devices can also reduce the problems caused by belching gas.

Compartmentalization. As indicated earlier, compartmentalization can enhance the operation of the septic tank. Some tanks are designed with multiple compartments, and some localities stipulate the use of two-compartment tanks. A two-compartment tank, as shown in Figure 7.2, helps to eliminate the possibility of short-circuiting wastewater through the system. Testing of one and two compartment tanks with similar wastewater streams showed that effluent levels of BOD (biological oxygen demand) and TSS (total suspended solids) levels were lower from the two-compartment tank than the one-compartment tank[8]. BOD can be reduced by 50 percent, and TSS can be reduced by as much as 80 percent by the introduction of a second compartment[9].

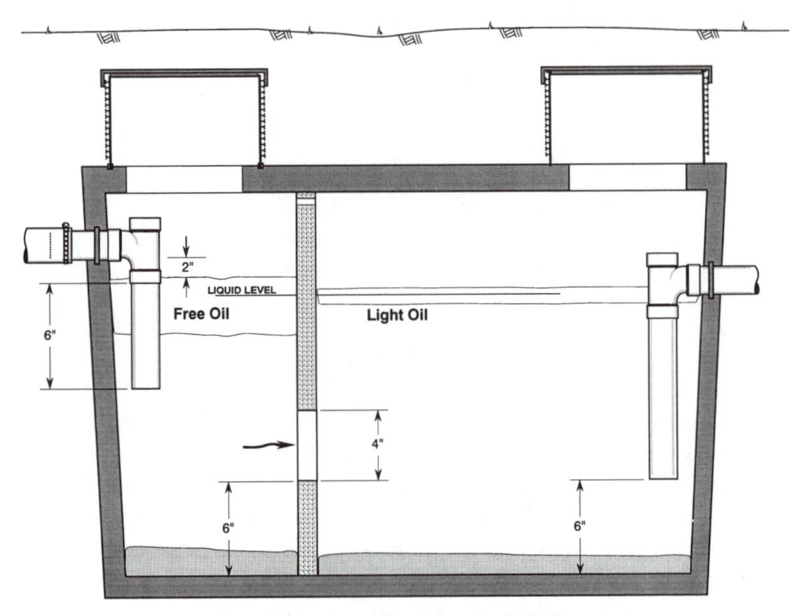

Figure 7.3 - Grease Trap

Grease Traps. Other designs use several tanks in a series, the first a grease trap designed to cool and capture all of the "grease" (actually, the oil, fat and grease) entrained in the wastewater. If the wastewater is quite warm, such as hot, greasy water from a diner, the grease cannot quickly separate from the water, so the grease flows right through the tank into the soil and congeals after cooling; the grease can fill the pores of the soil and ruin the drainage system. Therefore, grease traps are mandatory for restaurants and food service facilities so that the grease is allowed to separate from the wastewater by floating to the surface of the trap. Once there, the grease is prevented from leaving the trap by baffles and effluent discharge pipes that extend below the presumed depth of the oil and grease. Figure 7.3 shows a typical grease trap with devices to minimize grease spillage into the septic tank. When using a grease trap, the designer must determine whether or not the facility will be using emulsifiers — cleaning products that keep the oil and grease in suspension — since the use of those products may allow the oil and grease to short-circuit the grease trap.

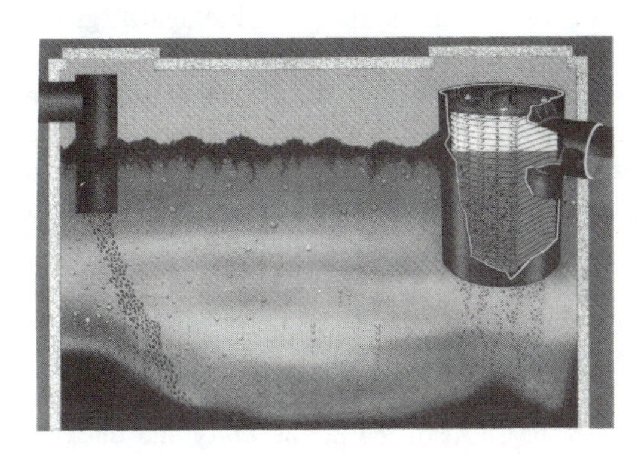

Figure 7.4 - Effluent Filter
(Courtesy of Zabel Wastewater Filter Systems, Louisville, KY)

In lieu of compartmentalization, some jurisdictions will allow the designer to use an effluent filter, similar to Figure 7.4, to strain the septic tank effluent. such filters can keep suspended solids to less than 2 mm (1/16 in) in size[10].

Soil absorption system. A typical soil absorption system consists of several components: an excavation (referred to as a "trench," "bed," or "cell," depending on its width and depth), gravel, distribution pipes, geotextile fabric, and topsoil. The excavation provides soil surface areas for the application of wastewater and soil medium to support biological growth, the major treatment process of the system. Wastewater passes through the soil pores of the bottom and sidewalls of the trench where the biological growth has affixed itself to the soil particles. Gravel supports the excavation walls, preventing them from collapsing onto the distribution pipe and bottom. Distribution pipes carry septic tank effluent to the soil absorption area and distribute it to the soil surface. Geotextile fabric protects the trench from the in-washing of "fines"— silts and clays — that can seal the pores. The topsoil protects the trench surface and geotextile fabric. Figure 7.5 shows the components of a typical trench in cross-section.

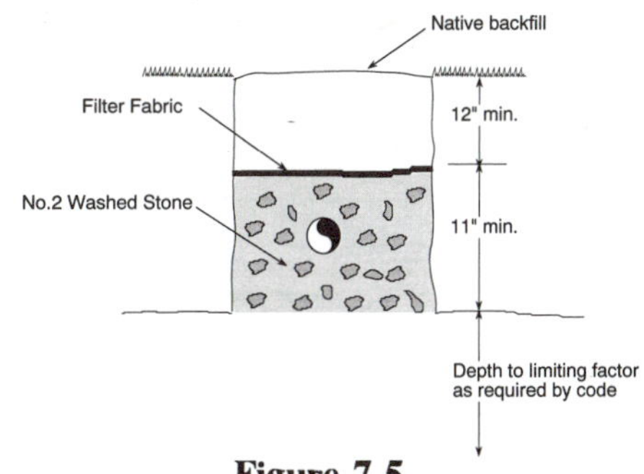

Figure 7.5
The components of a trench in cross section

Like any engineered device, an onsite system has a *design life*, which is the estimated length of time before the system will have to be replaced or rehabilitated[11]. For an onsite system, the critical factor is the soil absorption system. Soil is the least uniform component of the system and the most difficult to monitor for operational problems. Current designs often assume that the soil absorption area will eventually "fail." *Failure* is the inability of water to penetrate the soil. This impenetrability is often attributed to the biological medium that grows on the soil particles. Some scientists argue that this "failure" is simply the result of the maturation of the biological growth throughout the soil absorption area, "maturation" simply meaning that a biological mat (or *biomat*) has formed throughout the soil absorption area. When the biological mat is spread evenly throughout the soil absorption area, wastewater moves from the system to the soil at a constant rate. Designers may identify an alternate site on the property where a second soil absorption area can be installed if the first field fails[12].

Development of a Biomat. The biomat is formed at the point where the septic tank effluent enters the soil, as shown in Figure 7.6. The biomat is dynamic: it expands and shrinks with the changes in wastewater quality and quantity and with the growth and decay of microbes. Also, the permeability and thickness of the biomat is affected by the type of soil and other environmental factors.

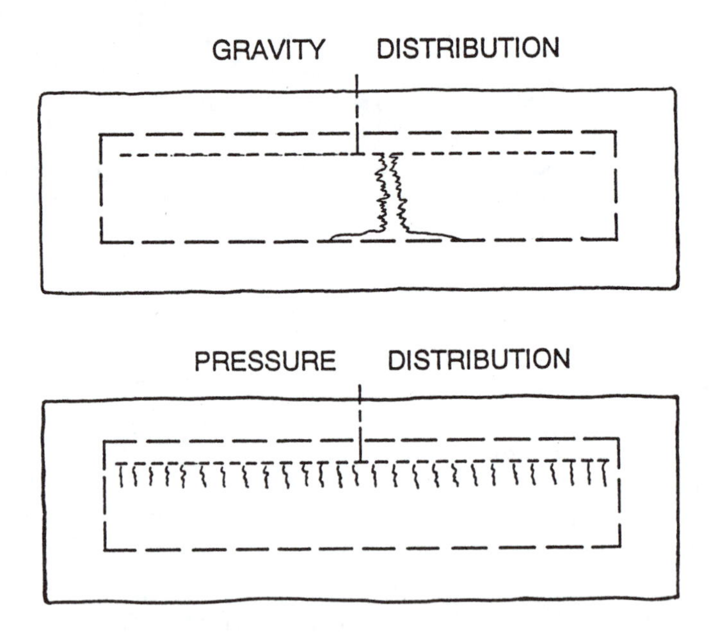

Figure 7.6 - Progressive Maturation

The layer is from 5 mm to 30 mm thick and composed of wastewater solids, minerals, microbes and cell hulls. The BOD_5 ranges from 3,000 to 42,000 mg/g of soil. (These concentrations are for quantities of soil, not liquid[13].) Biomats are highly effective in removing organic material and pathogens from wastewater.

A mature biomat has also been shown to detain viruses from the effluent[14]. A biomat has a low permeability, about 0.5 ft/day, which controls wastewater infiltration. The flow beyond the biomat occurs through capillary action, which further enhances treatment.

There are two general maturation modes for a biomat: serial and distributed, depending on how wastewater is transported and distributed in the soil absorption area. The maturation mode is a function of the distribution method. In gravity-fed systems, maturation occurs through "serial" growth, effluent not travelling further down the distribution pipe than the first few holes. Biological growth starts near the perforations where the effluent is discharged. As the system ages and the soil "clogs," the maturation surface moves along the length of the trench or trenches until the entire soil absorption area has a mature biological surface[15]. A passive gravity dosing system, such as the apparatus shown on page 172, promotes distributed maturation. This adaptation to a distribution box provides for more even distribution of effluent without the need for pumps.

In a pressure distribution system, the effluent is applied evenly throughout the soil absorption area. The biological growth is distributed evenly throughout the soil absorption area. The same processes are working, but occurring uniformly in the vicinity of each perforation. Eventually the biomat will mature throughout the whole distribution area.

The concept of progressive maturation is sometimes difficult to evaluate because of a reduction in *available infiltrative surface,* a phenomenon that is said to occur when "dirty" (with sand and clay) gravel in a soil absorption system prevents movement of water by clogging soil pores. The poor-quality gravel reduces the surface area between the wastewater and the soil. This reduction in infiltrative surface reduces the effective bottom and sidewall area for water movement, as shown in Figure 7.7. Whether or not this reduction occurs, experience supports siting an alternate field[17].

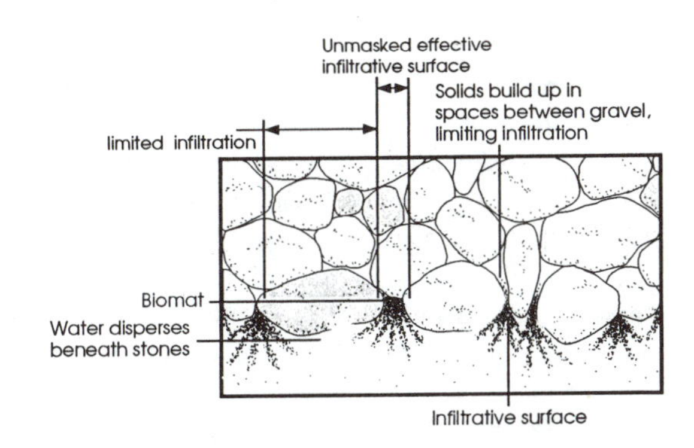

Figure 7.7 - Reduction in available infiltrative surface as a result of low quality gravel

It is important to understand the impact of the biological layer on the soil absorption system. The movement of septic tank effluent through a biological layer is a special case of the movement of water through a saturated soil. This movement, which is called *soil permeability,* can be estimated by using Darcy's

Laws, which are explained in Chapter 5, *Hydraulics*. The expression h_L/L is the *hydraulic gradient* and is sometimes known by the term i.

$$v = -k\left(\frac{h_L}{L}\right) \tag{7.2}$$

where
V = Velocity, m/min (ft/min)
h_L = Headloss, m (ft)
L = Length of flow, m (ft)
k = hydraulic conductivity, m/min (ft/min)

The Darcy Formula for flow, Equation 7-3, is simply the product of the velocity and the area of the cross section of the soil sample.

$$q = vA \tag{7.3}$$

where:
q = Flow, m^3/min (ft^3/min)
V = velocity, m/min (ft/min)
A = Cross sectional area of the sample, m^2 (ft^2)

Soil Absorption Area Design

To function properly — treat septic tank effluent and dispose of the water — a soil absorption system must be designed in accordance with the limitations of the site, the type of soil and the dispersal method. Foremost among these is the distance to a *limiting factor* — the condition that limits treatment capacity: groundwater or bedrock. Other limitations affecting a site include slope, selection of excavation width, geometry of the system, and layout of the site.

Vertical Separation. Vertical separation is a hotly contested issue throughout the nation. A 900 mm (36 in) separation is often suggested as a safe distance between an infiltration surface and a limiting factor. This separation provides at least 600 mm (24 in) of soil for treatment and an additional 300 mm (12 in) as a safety factor to account for the vagaries of the soil, such as root channels and rocks[18]. A 900 mm separation provides a high degree of protection, except in the coarsest of soils, if hydraulic loading rates are low. Research on sand filters demonstrated that sand depths of from 1,200 mm 1,500 mm (4 ft to 5 ft) are necessary for coarse sands[19].

Laboratory and field studies on septic tank effluent tend to confirm a 900 mm separation distance. Magdoff reported complete removal of bacteria in 900 mm soil columns of sand underlain by silt loam[20]. Others have shown significant bacterial removal within 300 mm soil columns and 600 mm beneath trenches.

Complete removal within the trenches occurred between 600 mm and 1200 mm (4 ft) beneath the trenches[21,22].

The 900 mm separation is not universally held; different states apply different standards. Appendix A lists the minimum separation distances for each state, which range from 200 mm to 1,800 mm (8 in to 72 in). The huge difference is due to many factors, none of which have anything to do with science. Some values result from traditions based on "what works" while others are based on local research or "worst case" situations. In some cases, the separation distance is even intended to limit development because it has the effect of severely restricting the lots that can be developed. Cogger identified three principle reasons guiding the development of vertical separation distances: varying interpretations of the research literature, interactions of hydraulic loading rate and soil type, and economic considerations[23].

Slope. Slope is another site limitation that must be considered. Generally, but not always, water beneath the surface flows parallel to the ground surface. Soil absorption trenches must be sited so that effluent will not break out downslope. Moreover, the soil absorption area should be sited downslope of any wells and beyond any well's zone of influence.

Excavations should be parallel to contours but not along convergences. Placing excavations along a slope provides the longest area for dispersal of the water. In contrast, an excavation placed parallel to a slope would quickly fail because all of the effluent would pond at the low end.

Excavation Width. Excavations are classified according to their width. Narrower excavations, those between one and five feet wide, are called *trenches*. Excavations wider than five feet are called *beds*. This designation based on width is somewhat arbitrary; some designers consider excavations to be trenches if they are from one to three feet wide.

Another way to distinguish trenches from beds is by the number of distribution pipes in the excavation. Trenches have only one distribution pipe in the excavation; beds contain multiple distribution pipes in a single excavation.

Beds tend to be installed at sites where space is scarce but are more prone to failure. The reason for this increased possibility of failure is from reduced oxygen transfer. In a trench, oxygen can move between the undisturbed sidewalls and the excavation[24]. However, in a bed the sidewalls are sometimes too far apart to provide sufficient oxygen for the entire excavation. Another factor is the surface area available on the sidewalls. Depending on the jurisdiction, trench designs are based on the surface area only of the bottom of the trench, the sidewall surface areas not usually factored into the design. (This is the result of attempts to provide "fail safe" designs through code requirements.) In a trench with six inches of sidewall surface area on each side, an extra square foot of absorption area is available for each foot of trench length. This is a 25 percent more infiltrative surface that is not accounted for in the design.

Trenches are effective only if the sidewalls do not interfere with each other. If trench-to-trench interference happens, the trench would behave like a bed. Many designers separate trenches by three feet and successfully avoid trench-to-trench interference.

Replacement Area. As an additional factor of safety, many designers map out a replacement soil absorption area in case the initial area fails. Some administrative rules mandate alternate sites as a part of the design approval process. The "replacement" can take several forms, depending on the space available, the characteristics of the soil, and the ingenuity of the designer. For example, if sufficient room exists, an entirely separate soil absorption area can be designated and protected. If space is limited, some designers will use the area between trenches as the replacement system and two alternating sets of pipes can be installed. Or, for small lots where soil conditions are suitable, the replacement area can actually be the same site but deeper in the soil.

Discharge. The soil absorption system can receive effluent from the septic tank in one of three methods: gravity, dosing, or pressure distribution. The *gravity-fed* system is the simplest. Because the septic tank has a fixed capacity and is always full, any new wastewater forces an equal volume of the contents out of the septic tank and into the distribution pipe where it flows by gravity to the soil absorption area. Typically, the distribution system consists of a distribution box (or "D-Box") which splits the flow equally among the laterals, which are usually constructed of 100 mm (4 in) perforated PVC pipe.

Dosing is the second method of discharging effluent to the soil absorption area. Dosing entails periodic discharges to the soil. The period can be based on when the dosing tank fills, or on a time basis. Dosing uses the principle of *dosing and resting*. Under this principle, the soil absorption system works more efficiently because it receives effluent at specific intervals that ensure that the soil will have time to "rest" or "recharge" between doses. This so-called resting is actually just a period of unsaturated flow through the soil rather than no flow[25].

Dosing can be achieved by a pump or siphon. Siphons are used because they require no energy and, in theory, work indefinitely if they are properly installed and maintained. In practice, however, siphons may fail because they leak or become plugged. Pumps provide a more reliable dosing method. When connected to a timer, pumps can dose the soil absorption area in periods and doses specified by the designer. Dosed systems are more like gravity-fed systems in that the effluent is delivered to the distribution box, and gravity is used to distribute the wastewater in the laterals.

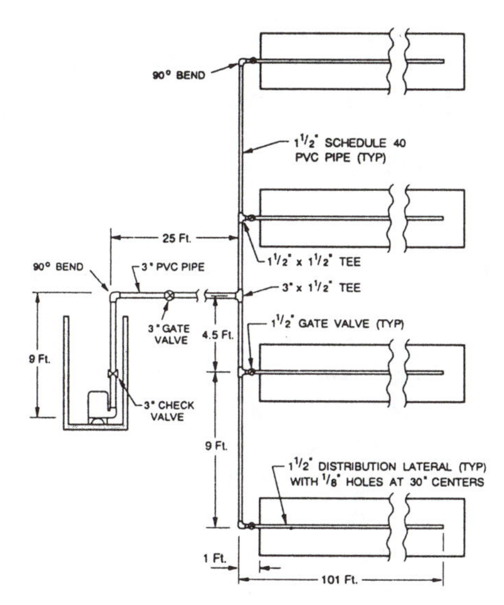

Figure 7.8 - Pressure Distribution System

Pressure distribution, as shown in Figure 7.8, is the third method of delivering effluent to the soil absorption system. The major difference between pressure dosing and pressure distribution is that pressure distribution relies on fully pressurizing the distribution pipe, which may be as little as 34 mm (1-¼ in) diameter and have distribution holes 8 mm (¼ in) in diameter. The combination of full pressurization, small pipe diameter, and small perforations ensure that effluent will be discharged equally along the lateral.

Design Examples

It is helpful to examine typical installation designs and compare them to each other. Some examples are included here for pressurized and non-pressurized designs. Each example is intended to show the various considerations that must be taken into account during the design process.

Septic Tank — single distribution line.

An owner intends to build a four-bedroom house and needs a septic system. The soil report indicates that wastewater can be applied to the soil at the rate of 0.5 gal/day-ft^2. The local sanitation code specifies a flow of 150 gal/day per bedroom. The code also specifies that trenches can be no wider than three feet, and only the bottom area can be considered as an infiltrative surface. Design a simple system.

Tank size: Based on the code, the house will have a code specified flow of 4 x 150 = 600 gal/day. Using a detention time of 36 hours in the septic tank, the capacity of the tank will be, according to Equation 7.1, (36/24) x 600 = 900 gallons. The local septic tank manufacturers supply 1,000 gallon tanks. Therefore, the septic tank capacity will be 1,000 gallons.

Soil absorption area: At a loading rate of 0.5 gal/day-ft^2, the required soil absorption area will be 600/0.5 = 1,200 ft^2. At a maximum width of 3 feet, the total length will be 1,200/3 = 400 feet.

Special Conditions: The local code recognizes the presence of gravel as reducing the available infiltrative surface and allows a 40 percent reduction if leaching chambers are used. The designer specifies a leaching chamber that has a two-foot width. The total bottom area required is (1-.4) x 1,200 = 720 ft^2. The total length is 720/2 = 360 feet.

Variances: The designer petitions the local sanitarian to use the sidewall area beneath the distribution pipe. According to the local code, the distribution pipe must be bedded 6 inches above the infiltrative surface. If successful, the designer will have 2 x 0.5 = 1 ft of additional infiltrative surface for each foot of surface area. The total length will be 1,200/(3+1) = 300 feet.

Septic Tank — multiple distribution line. The owner decides to remove from consideration part of the area originally set aside for the single distribution pipe. The designer must now install multiple laterals, each receiving equal flow. Three laterals will be installed instead of one. The local sanitation code requires that each trench be separated from adjacent trenches by three feet. The petition to use the sidewall area was granted, so three 100-foot trenches are installed. The width of the soil absorption area is (3 x 4) + (2 x 3) = 18 feet. The laterals are connected to a distribution box that has three outlets. Each outlet is set to the same elevation so that each lateral receives equal flow.

Septic Tank — Dosed System.
Based on a discussion with the owner, the designer learns that all of the flow will be generated in the morning and evening. To equalize the flow, a dosing chamber will be installed to pump equal doses 4 times daily. In this example, a dose would be (600 gal/day)/(4 doses/day) = 150 gal/dose. The designer selects a dosing tank having a capacity of 125 gallons per foot. If the designer

doses based on volume, float switches will have to be set (150/125) x 1 ft = 1.20 ft apart. If dosing is based on time, the timer will have to be set to activate once every six hours.

Pressure Distribution System: The designer decides to go with a pressure distribution system. He will install four 3-foot trenches, each 100 feet long, having a total bottom area of 1,200 square feet. By code, the trench walls must be six feet apart. All trenches will be installed at the same elevation. Each lateral will be fitted with a cleanout at its end and a regulating plug valve at the lateral head. In addition, the header will be fitted with a plug valve, and swing check valve will be installed to protect the pipe. According to code, the distal residual pressure must be at least 2.5 ft of head. The designer will use Schedule 40 PVC pipe, which has a C_v of 150.

The designer wants a minimum water velocity of 2 ft/sec and a hole spacing of 2 ft. He initially selects 1½ inch pipe and ⅛ inch holes. Using Equation 5.26, the flow per orifice is equal to 11.797 $(.125)^2$ $(2.5)^{.5}$ and 0.296 gpm. The lateral will have 50 holes; therefore the flow will be 50 * 0.296 = 14.572 gpm/lateral. The designer must now make sure that the minimum velocity in the lateral is 2 ft/sec. This can be determined by using the following:

$$V = 0.409\left(\frac{Q}{d^2}\right) \qquad (7.4)$$

where
- V = Velocity, ft/sec
- Q = Flow, gpm
- d = Diameter, in

In this case, the velocity is equal to 0.409 * (14.572) * $(1.5)^2$ and equals 2.646 ft/sec. Had the designer selected a different hole spacing, she could determine the number of holes in the distribution pipe by using one of the two equations listed below, depending on whether the manifold is located centrally or at one end of the distribution pipe.

$$N = 0.5 + \left(\frac{P}{X}\right) \qquad (7.5)$$

where:
- N = The number of laterals
- P = Lateral Length, ft
- X = Hole Spacing, ft

For an end manifold, Equation 7.5 becomes:

$$N = 1 + \left(\frac{P}{X}\right) \tag{7.6}$$

The headloss per lateral is calculated from Equation 5.11 and equals 1.94 ft.

$$h_L = 0.002083\ (100) \left(\frac{100}{150}\right)^{1.85} \left(\frac{(14.572)^{1.85}}{(1.5)^{4.8655}}\right) = 1.94\, ft \tag{7.7}$$

In a like manner, the headlosses in each pipe segment are calculated. For the 9-foot segments, the headloss is calculated the same way. In this problem, the headloss is 0.175 ft. In the 4-foot segment, the flow is doubled because it feeds two laterals, and when calculated, is 0.315 ft. The headloss in the other two laterals and feeder pipe are identical.

The *header pipe* must now be selected. The header pipe must carry the entire flow, which is 58.288 gpm. The velocity in the pipe must not be below 2 ft/sec and should not exceed 5 ft/sec. At higher velocities, the headloss will increase exponentially and result in both excessively large pumps and wear on the system. (In this example the velocity exceeds 5 ft/sec but only for 4.5 feet and is not significant.) Using Equation 7.4 and trying a 3-inch Schedule 40 PVC pipe, the velocity is calculated to be 2.646 ft/sec, which is acceptable. The 3-inch pipe is a total of 34 feet, 25 feet of header pipe and nine feet inside the pump chamber. The total headloss for the 3-inch pipe is .226 feet.

Headlosses for fittings and valves must now be calculated. While headloss in an individual fitting may seem small, the cumulative headloss can significantly add to the system losses. A conservative approach allows 31 percent for headlosses in valves and fitting if the losses are not calculated individually and is suitable in most cases[26]. In this example, these headlosses have been calculated to illustrate their significance and for comparison to the recommendation. Headlosses will be calculated according to Equation 5-8 and are summarized in Table 7.2.

In the preceding example, the headloss has been calculated using C_v equals 150, which is a standard value for new PVC pipe. In practice, a distribution pipe has a design life of 20 years during which time the pipe will be subject to wear, clogs, and debris, all the result of repeated use. After 20 years of use, a more appropriate value for C_v is 140. If that is the case, a different pump might be considered because the performance of the system will change by seven percent. The actual decision would be based on the operating characteristics of the pump the designer has under consideration. Designers take this into account when developing a system curve, and they will draw two curves, one for each C_v value, as discussed in Chapter 5, *Hydraulics*.

Table 7.2 - Headloss Values				
Fitting	Size(in)	Number	K value	Headloss (ft)
Check Valve	3	1	1.70	0.185
Gate Valve	3	1	0.14	0.015
90° Bend	3	1	0.51	0.055
Tee, Branch	3 X 1½	2	1.26	0.548
Tee, Branch Tee, Flow	1½ X 1½	2 2	0.42 1.42	0.046 0.548
Gate Valve	1½	4	0.14	0.015
90° Bend	1½	2	0.51	0.055

When the pressure distribution system is put into operation, the flow to each lateral will have to be equalized. There are different methods for ensuring equalized flow. For example, the hole spacing and length of each lateral could be calculated based on the system head at the manifold. In this example, valves are installed at each header, and fittings for standpipes will be installed at the end of each lateral. When the pressure distribution system is installed, the flow to each lateral and residual pressure can be measured and calibrated directly. As the system ages, each lateral can be recalibrated to assure that it receives and distributes the flow as the designer intended.

Using valves and standpipes result in standardized designs for each lateral. Specifying customized laterals with no standpipes forces the designer to assume that the system is operating as it was designed. The designer cannot verify the accuracy of the design or correct for minor errors resulting from material imperfections, installation techniques or miscalculations. Valves and standpipes allow the designer to examine the system throughout its lifetime and adjust it for optimal performance.

Dose Volume. To function correctly, a pressure distribution system must be completely filled with wastewater and operating at its design head. Then, it must deliver the volume of wastewater specified by the designer. The designer must determine how much wastewater is necessary just to fill the pipes. To this volume, the designer must add the dose of wastewater that will be applied to the soil. Keep in mind that when the pump shuts off, the pipes will be full of wastewater. Some of this water will dribble out of the orifices, but most of it will remain in the pipe. Because water in pipes can freeze or deposit fine suspended solids, it is better to have this wastewater drain back into the pump chamber.

Conservative designers recommend that the dose volume be between five to ten times the lateral volume. In this example, the volume of the pipe is equal to 400 feet of 1½ inch pipe for a total of 36.72 gallons. There is also 26 feet of 1½ inch manifold that has a volume of 2.29 gallons and 34 feet of 3 inch of forcemain with a volume of 12.48 gallons. The volume of pipe to be filled is 51.49 gallons, and the minimum dose volume is 183.60 gallons. Because the house is designed to have a daily flow of 600 gallons, the designer can provide for 3 daily doses of 200 gallons each for a total of 600 gallons per day, for a dose volume that is 5.45 times the lateral volume.

Pressure Distribution Systems on Sloping Sites

The previous example was relatively simple to design because the laterals were all at the same elevation. In actual practice, many systems are installed on sloping sites. Because of slope, the laterals are placed at different elevations and will be subject to different total heads. There are alternative control strategies a designer can use to provide equal distribution despite the different elevations, each of which is intended to provide for wastewater application at the design loading rate:

- Regulating Valves
- Differential Hole Spacing
- Differential Hole Sizing
- Differential Lateral Lengths

Consider a pressure distribution system with laterals at elevations of 100 Mean Sea Level (MSL) and 95 MSL. Because of the 5-foot elevation difference between the two laterals, the residual head at the lower elevation will be at least 7.5 feet if the residual head of 2.5 feet at the higher elevation is to be met. The designer prefers to install four laterals branching off of a center manifold. For the higher elevation, the laterals will be 75 feet long each and use ⅛-inch holes on two foot centers. Using Equation 5.4, the total flow to each lateral will be 11.25 gallons per minute, for a total of 22.50 gpm to the upper laterals. To provide equal distribution, each of the lower laterals will receive 11.25 gpm each also.

To calculate the correct hole spacing and hole size, the system losses must be calculated. First, the headloss in the 25 foot manifold is calculated. Then the headloss in each lower lateral is calculated. These headlosses are 1.09 ft and 0.95 ft, respectively, for a total of 2.04 ft. Because the total head will be over twice as high, the designer should select a hole spacing corresponding to the higher head. The hole spacing will be 5.25 feet. If the lower manifolds are 75 feet long and using Equation, 7.6, there will be 15 holes in the lower manifolds.

Rearranging Equation 5.26 and calculating with 15 holes and a residual head of 6.41 feet, the orifice diameter will be .164 inches, which is nonstandard. The designer selects 5/32 inch orifice openings and recalculates the number of holes necessary to achieve 11.25 gpm. The answer is 16.46 holes, so the designer selects 17 holes, which according to Equation 7.6, will necessitate a lateral 89.25 feet long. Using that length and 17 5/32 holes, the total flow out of the lower laterals is 11.42 gpm, which is 1.5 percent higher than the flow out of the higher elevations and within the tolerance of the formulas used to calculate the flows and headlosses.

As an alternative, the designer could have elected to install regulating valves on the lateral to the lower laterals. While this would have simplified calculations, it would cause unnecessary wear on the pump. Or, the designer could have installed an orifice plate or used a different diameter pipe.

The more important question is whether the lower lateral is receiving the same quantity of wastewater as the higher lateral. True, all laterals are receiving nearly identical flows, but the lateral at the higher elevation has more orifices, each distributing flow at a lower total head. The two laterals at the lower elevation have holes that are spaced further apart and discharging head at a higher total head. For truly uniform distribution throughout the soil absorption system, the designer would argue for the use of regulating valves.

WISCONSIN MOUND SYSTEM

The Wisconsin Mound System (or simply the "mound" system) is a single-pass sand filter intended to provide both treatment and subsurface disposal of the wastewater. It was developed in the 1970's to address the limitations of "conventional" technologies available at the time, and the stated goal of the designers was "to treat and dispose of the wastewater via the subsurface in an environmentally acceptable manner and to protect the public health[27]."

The mound system consists of four basic components: septic tank, pump chamber, pressure distribution system, and sand fill. The mound itself will be described in accordance with the conventions used for its design and installation. Figure 7.9 shows the components of the mound.

The siting and design of a mound is based on two critical factors: how effluent will disperse from the mound and the rate for the dispersal. A mound is designed on the assumption that the wastewater applied to it will be "typical domestic sewage." If the wastewater strength exceeds those values, (see Table 2.1 for wastewater concentrations) alternative measures may be necessary.

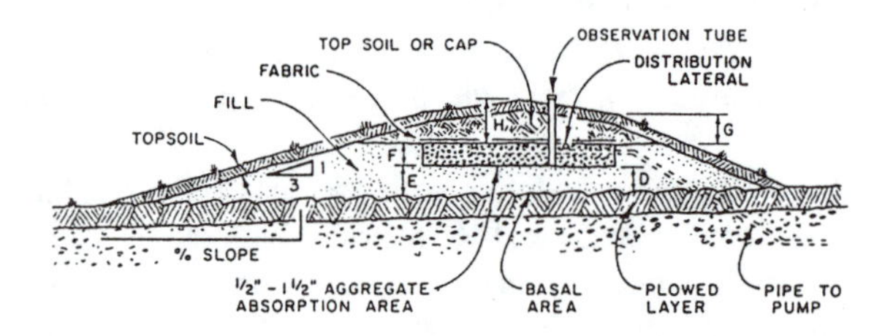

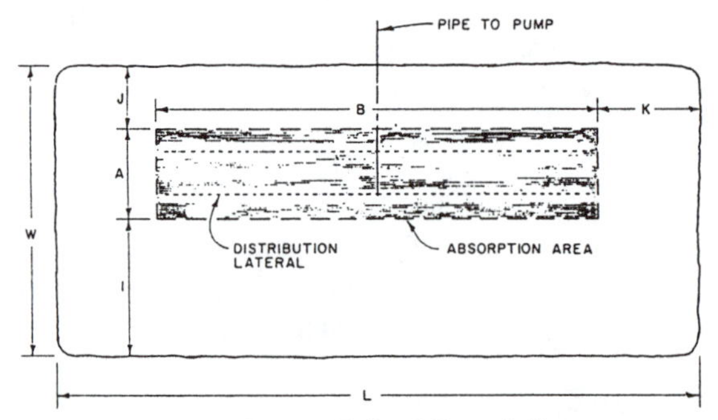

Figure 7.9 - Mound System

There are three critical loading rates that must be successfully estimated if the mound system is to function properly. These are:

- the sand fill loading rate,
- soil loading rate (which is also called the basal loading rate), and
- the linear loading rate.

The *sand loading rate* is the rate at which the wastewater is pumped into the mound. By convention, the sand loading rate is 1.0 gal/day-ft^2. The *basal loading rate* is based on the soil infilatration rate at the sand-soil interface beneath the mound. Table 7.3 lists basal loading rates for typical soil conditions. Once these values have been established, the mound design can proceed.

The *linear loading rate* is a more sublime function of the water movement, somewhat harder to predict. Linear loading refers to the unit volume of water discharged along each foot of mound length each day. Linear loading is critical to understand because a mound will leak at its *toe* (downslope side) if it is overloaded along its length.

The linear loading rate varies, depending on the direction of water movement away from the mound. On level sites and where water movement is essentially vertical, the linear loading rate can be high, from 8 gal/day/lf to 10 gal/day/lf.

On sloping sites where water movement is essentially horizontal and shallow groundwater exists, and in platy soils, the linear loading rate should be from 3 gal/day/lf to 4 gal/day/lf.

The dimensions of a mound are identified according to the *Wisconsin Mound Soil Absorption System Siting, Design and Construction Manual*, which is produced by the University of Wisconsin-Small Scale Waste Management Project[28]. These dimensions, each of which are identified by a letter, are identified below and shown in Figure 7.9.

A Soil Absorption Area Width: The width of the sand fill area.

B Soil Absorption Area Length: The length of the sand fill area.

D Sand Fill Depth: The minimum depth of sand at the **upslope** end of the mound to meet local separation distances. The depth of sand fill plus native soil must be equal to the required separation distance.

E Sand Fill Depth: The minimum depth of sand at the **downslope** end of the mound. Because the sand fill must be level beneath the distribution system, E is equal to D plus additional depth to keep the distribution system level.

F Distribution System Depth: The distribution system must be at least 9 inches deep because there must be at least six inches of aggregate below the distribution pipe, the distribution pipe diameter (one inch diameter minimum) and two inches of aggregate cover above the distribution piping.

G Cover Depth on the Upslope Edge of the Distribution System: The upslope cover is a minimum of six inches in warmer climates and 12 inches for colder climates.

H Cover Depth at the Top of the Mound: The mound should have a minimum of 12 inches of cover for warmer climates and 18 inches of cover for colder climates.

I Slope Width of Downslope Side of the Mound: Basal width on the downslope side of the mound.

J Slope Width of Upslope Side of the Mound: Basal width on the up-slope side of the mound. Note: For mounds on level sites, I and J will be equal. For sloping sites, I will always be longer than J because the water flow, and hence absorption area, will be in the direction of I.

K Slope Length: The length of the sloped sides of the mound, based on a 3:1 slope from the top of the mound. A typical dimension is 10 to 15 feet. Steeper slopes are not recommended because they may pose a safety hazard but are acceptable if construction practices take into account the inherent dangers of steeper slopes.

L Length: The length of the mound from endslope to endslope.

W Width: The width of the mound from endslope to endslope.

Table 7.3 - Basal Loading Rates		
Soil Condition[1]	**Beds** Gal/ft² (L/m²)	**Trenches** Gal/ft² (L/m²)
Is the soil texture of the entire profile 3' below the infiltrative surface extremely gravelly sand, gravelly coarse sand or coarser?	0.4 (16.25)	0.4 (16.25)
Is the soil structure of the horizon moderate or strong platy?	0	0.2 (8.13)
Is the soil texture of the horizon sandy clay loam, clay loam, silty clay loam, silt loam or finer, and the soil structure weak platy?	0	0.3 (12.2)
Is the moist soil consistence of the horizon stronger than firm or any cemented class?	0	0
Is the soil texture of the horizon sandy clay, clay or silty clay of high clay content, and the soil structure massive or weak?	0	0
Is the soil texture of the horizon sandy clay loam, clay loam, silty clay loam, silt or silt loam and the soil structure massive?	0	0.2 (8.13)
Is the soil texture of the horizon sandy clay, clay or silty clay of low clay content and the soil structure moderate or strong?	0.2 (8.13)	0.3 (12.2)
Is the soil texture of the horizon sandy clay loam, clay loam, silty clay loam or silt loam and the soil structure weak?	0.2 (8.13)	0.3 (12.2)
Is the soil texture of the horizon sandy clay loam, clay loam or silty clay loam, and the soil structure moderate or strong?	0.4 (16.25)	0.5 (20.4)
Is the soil texture of the horizon loam or sandy loam and the soil structure massive?	0.3 (12.2)	0.4 (16.25)
Is the soil structure of the horizon loam or sandy loam and the soil structure weak?	0.4 (16.25)	0.5 (20.4)
Is the soil texture of the horizon sandy loam, loam or silt loam and the soil structure moderate or strong?	0.5 (20.4)	0.6 (24.4)
Is the soil texture of the horizon very fine sand or loamy very fine sand? Or does it have the structure as the next category, but with a massive soil structure?	0.4 (16.25)	0.5 (20.4)
Is the soil texture of the horizon fine sand or loamy fine sand?	0.5 (20.4)	0.6 (24.4)
Is the soil texture of the horizon loamy sand, sand or coarse sand?	0.7 (28.5)	0.8 (32.6)

[1]If the answer to any of the Soil Condition descriptions is "yes," then the infiltrative, exposed natural soil surface for the system should be sized using the recommended soil loading factors.

The mound system has factors of safety built into it to account for the variation of uses to which it might be subjected. Specifically, the mound system for a residence is designed under the assumption that the daily flow per bedroom will be 150 gallons, 75 gal/day per occupant and two occupants per bedroom. Studies have shown, however, that daily water usage in rural residences is closer to 50 gal/day per occupant. A mound system design may be altered if the flow is known, but if no factors of safety are considered, the mound could quickly fail from organic overloading.

Design Example

Design a mound system for soil, site, and sizing conditions listed below and using the design specifications outlined in the previous section.

Use: Four-Bedroom house designed for a flow of 150 gal/day-br

Soil: Sandy loam having weak structure
Estimated high groundwater at 26 in

Site: 10 percent slope
Horizontal soil water movement
Northern climate

Establish the Loading rates for the sand and soil:

For a sandy loam having weak structure, the basal loading rate is 0.4 gal/day-ft^2.

The soil water movement is horizontal, so select a linear loading rate of 3 gal/day-lf.

Select sand fill conforming to ASTM C 33 and use a sand fill loading rate of 1.0 gal/day-ft^2.

Calculate **A**

$A = (3.0 \text{ gal/day-lf}) \div (1.0 \text{ gal/day-ft}^2)$

A = 3 ft

Calculate **B**

$B = (600 \text{ gal/day}) \div (3 \text{ gal/day-lf})$

B = 200 ft

Calculate the Basal Width, (**A + I**)

$A + I = (3 \text{ gal/day-lf}) \div (0.4 \text{ gal/day-ft}^2)$

$A + I = 7.5$ ft

I = 7.5 ft - A = **4.5 ft**

Calculate **D**

The local code requires a minimum of 36 in of suitable soil. There is 26 in of suitable soil at the site. The minimum sand fill depth is 36 in - 26 in = 10 in of sand fill. The minimum depth of sand fill is 12 in, so **D = 12 in.**

Calculate **E**

The site has a 10 percent slope, so sand fill depth must take the slope into account if the bottom of the absorption area is to be level.

E = 12" + (0.10)(36")

E = 15.6 inches

Calculate Depths **F**, **G**, and **H**

F is based on there being a 6 inch aggregate base, 2 in PVC pipe, and 1 in aggregate cover. **F = 9 in.**

G = 12 in

H = 18 in

Calculate **J** assuming a 3:1 slope

J = [(3)(12 + 9 + 12)] in ÷ 12 in/ft

J = 8.25 ft

Calculate K, assuming a 3:1 slope

K = [(3)(½(12 + 15.6) + 9 + 18)] in ÷ 12 in/ft

K = 10.2 ft

Calculate **I** using a 3:1 slope and compare this value with the previous value

I = [(3)(15.6 + 9 + 12)] in ÷ 12 in/ft

I = 9.15 ft and is recommended over 4.5 ft

Overall Length, **L**

L = 200 feet + (2)(10.2)feet

L = 220.4 feet

Overall Width, **W**

W = 3 feet + 8.25 feet + 9.15 feet

W = 20.4 feet

Having laid out the dimensions of the mound, the designer is ready to design a pressure distribution system using the principles developed in Chapter 5. When the design is approved, and the system is ready for installation, the installer can refer to the installation instructions outlined in Chapter 9, Installation of an Onsite Wastewater Treatment System.

WISCONSIN AT-GRADE SYSTEM

The Wisconsin At-Grade System was designed for sites with soils too deep to justify a mound but too shallow to install a below-grade soil absorption system. The name of the system reflects its principal component: the distribution system is placed on the surface grade of the soil.

As with the design of a mound system, the designer must estimate the soil infiltration rate and predict the movement of the effluent from the system. The main component of the At-Grade system is the "effective absorption area," which is the soil area available to accept and transmit the effluent. This area is equal to the product of the length of the aggregate area and the distance from the distribution pipe to the toe of the aggregate area. Figure 7.10 details the components of an At-Grade system.

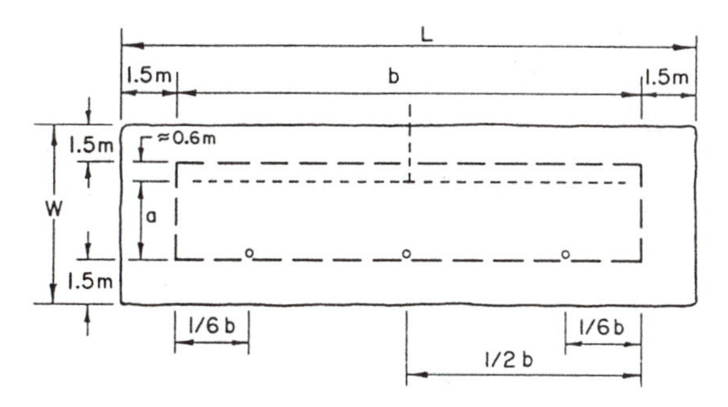

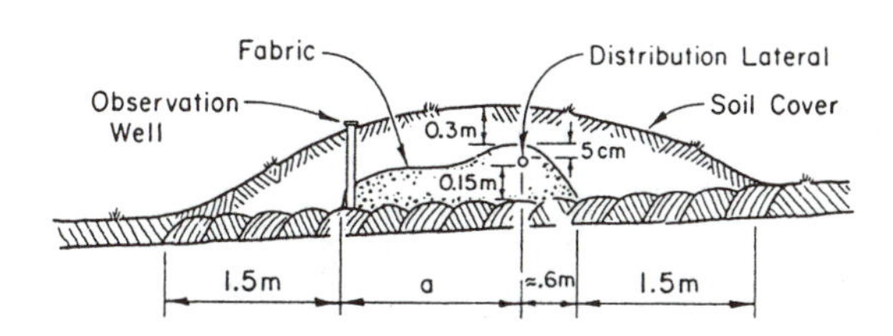

Figure 7.10 - At-Grade System

As with the mound system, Table 7.3 can be used to estimate the basal loading rate of an At-Grade system. Moreover, the linear loading rate of an At-Grade system is identical to that of a mound system.

Unlike a mound system, gravity distribution can be used to dose an At-Grade system. Gravity distribution is not preferred, but research conducted by the University of Wisconsin-Small Scale Waste Management Project failed to find evidence that gravity distribution is detrimental to an At-Grade.

Observation tubes are critical for an At-Grade system. The system maintainer can evaluate the condition of the At-Grade system based on evidence of ponding within the system. Observation tubes must be installed on the downslope edge of the aggregate.

Design Example

Design an at-grade system for soil, site, and sizing conditions listed below and using the design specifications outlined by Converse and Tyler. (Note: these are the same general conditions as the design example for the mound system except for the depth to high groundwater.)

Use: Four-Bedroom house designed for a flow of 150 gal/day-br

Soil: Sandy loam having weak structure
Estimated high groundwater at 36 in

Site: 10 percent slope
Horizontal soil water movement
Northern climate

Establish the loading rates for the soil

The soil loading rate from Table 7.3 is 0.4 gal/day-ft^2

The linear loading rate is 3 gal/day-lf

Calculate **A**

A = (3 gal/day-lf) ÷ (0.4 gal/day-ft^2)

A = 7.5 ft

Calculate **B**

B = (600 gal/day) ÷ (3 gal/day-lf)

B = 200 ft

Calculate **L** assuming 5 ft of cover at each end

L = 200 ft + (2)(5) ft

L = 210 ft

Calculate **W** assuming 5 ft of cover on each side and top, and **A**

W = 7.5 ft + 2 ft + (2)(5) ft

W = 19.5 ft

Like a mound system, an at-grade plan is executed and approved before the system is installed. The installation then proceeds in accordance with the specific installation requirements for an at-grade system.

SAND FILTER

Sand filtration, as discussed in Chapter 1, has been in use in the United States for over 100 years. Sand filters are quite effective, achieving coliform reductions exceeding 99 percent, BOD_5 removals of 97 percent, and TSS removal of 88 percent[29]. Sand filters can be installed above the ground, as is the case with the Wisconsin Mound, or they can be buried beneath the ground surface. A sand filter need not be installed in a tank. A lined excavation can serve as a "tank." Figure 7.11 shows the components of a sand filter.

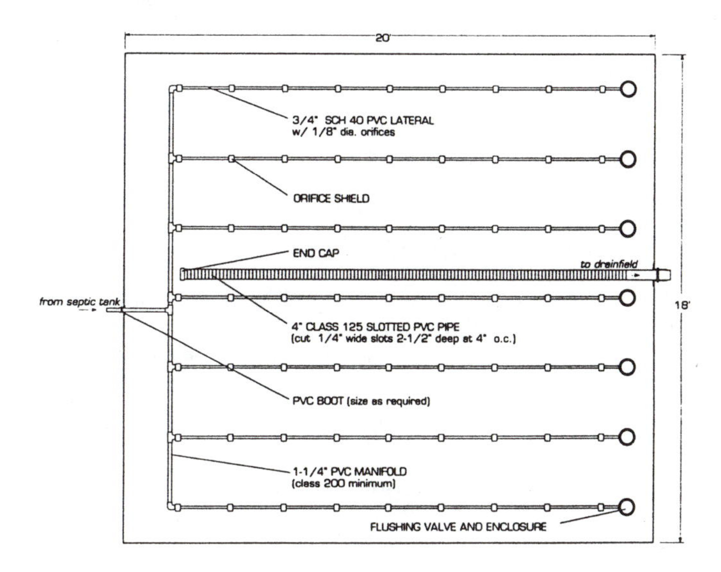

Figure 7.11 - Sand Filter

The performance of a sand filter is highly dependent on two factors: the grading of the sand media and the hydraulic loading rate[30]. At hydraulic loading rates of 0.2 m/day (5 gal/ft^2-day) and sand having an *effective size* of 0.19 mm and a *uniformity coefficient* of 3.31, septic tank effluent clogged an intermittent sand filters in three to five months but produced an effluent that met surface water discharge standards and which completely nitrified the septic tank effluent. An aerobic treatment unit effluent discharged to a sand filter at 0.14 m/day (3.5 gal/day-ft_2) required maintenance after nine months[31].

Dosing cycles are also critical to the success of the operation. Cycles have varied from once per day to once every 30 minutes. A sand filter should be dosed at least twice per day[32].

The quality of the effluent from a sand filter affects the method of discharge. As noted before, surface water discharge may be possible if nutrient loading is not a consideration. If surface water discharge is prohibited, subsurface infiltration may be possible at higher soil loading rates. Laak reports that reductions in soil absorption areas from 20 percent to 70 percent[33].

Design Example

Design an intermittent sand filter and soil absorption system. (Note: these are the same general conditions as the design example for the mound system except for the depth to high groundwater.)

Use: Four-Bedroom house designed for a flow of 150 gal/day-br

Soil: Sandy loam having weak structure
Estimated high groundwater at 36 in

Site: 10 percent slope

Select a dosing cycle and hydraulic loading rate:

The sand filter will be dosed with septic tank effluent four times daily and a hydraulic loading rate of 5 gal/day-ft^2.

Note: A pressure distribution system for the sand filter can be designed using the principles outlined in Chapter 5.

Calculate the sand filter area:

Area = (600 gal/day) ÷ (5 gal/day-ft^2)

Area = 120 ft^2

Select 10 ft X 12 ft tank, which is a lined excavation.

Calculate the soil absorption area and total trench length assuming a 50 percent reduction in area due to the high quality effluent discharged from the sand filter:

Based on Table 7.3, the soil loading rate for septic tank effluent is 0.4 gal/ft^2-day. For the sand filter effluent, the soil loading rate is 0.8 gal/ft^2-day. Select 3 ft wide trenches.

Area = (600 gal/day) ÷ (0.8 gal/day-ft^2)

Area = 750 ft^2

Total trench length = 750 ft^2 ÷ 3 ft width

Total trench length = 250 ft

SUMMARY

Presented in Chapter 7 are the most common alternative onsite systems constructed in the United States and Canada. Each has been extensively used and in a variety of climates, topographies, and soil conditions. There are local variations of all, depending on local codes and past practices of the locality, state, provence, or region. In addition, local factors such as groundwater elevation, topography, climate, and regional soils will affect the final design of any onsite system.

Alternatives mentioned in Chapter 6 have not been explored further in Chapter 7. There are several reasons for this, including limited use or extensive need for site-specific information. For example, an aquatic system is custom designed for each specific application; there is no general guideline that can be presented. Likewise, the selection of evapotranspiration is based on a number of local factors including pan evaporation rates, vegetation, and climate. The lack of a design example in Chapter 7 should not be construed to imply that these or other alternatives should not be considered if conditions for such a system are met.

References

1. Metcalf, Leonard. Transactions of the American Society of Civil Engineers, No. 909. December 1901. p. 472.

2. Salvato, Joseph. *Environmental Engineering and Sanitation.* New York: John Wiley and Sons. p. 411.

3. Laak, Rein. "Multichamber Septic Tanks," *J. Environ. Eng. Div., ASCE,* June 1980, pp. 539-546.

4 From "Soil Facts - Septic Systems and Their Maintenance" Michael T. Hoover, North Carolina Agricultural Extension Service, AG-439-138/90.
Source: Adapted from "Estimated Septic Tank Pumping Frequency,"
by Karen Mancl, 1984, *Journal of Environmental Engineering*. Volume 110.

5. Baumann, E.R., E.E. Jones, W.M. Jakubowski and M.C. Nottingham. "Septic Tanks." *Home Sewage Treatment.* Proceedings of the Second National Home Treatment Symposium. ASAE Publications, St. Joseph, MO, 1978.

6. United States Environmental Protection Agency. *Design Manual: Onsite Wastewater and Treatment and Disposal Systems.* Washington DC: Office of Water Program Operations. October 1980, p.99

7. Boyle, James, Rock, C.A., Edited by Robert W. Seabloom. "Performance of Septic Tanks." Proceedings of the 7th Northwest Onsite Wastewater Treatment Short Course and Equipment Exhibition. University of Washington College of Engineering, 1992, pp. 14-15.

8. Winneberger, J. *Septic-Tank Systems, A Consultant's Toolkit.* Volume II, The Septic Tank. Butterworth Publishers. Stoneham, MA, 1984. pp. 50-54.

9. Laak, R. "Multichamber Septic Tanks." *ASCE Journal of the Env. Eng. Div.*, June 1980, pp. 539-545.

10. Florida Administrative Code

11. Laak, R. *Wastewater Engineering Design for Unsewered Areas.* Lancaster: Technomic Publishing Co., 1986. pp. 86-88

12. The Wisconsin Administrative Code, Chapter ILHR 83

13. Laak, Rein. *Wastewater Engineering Design for Unsewered Areas.* Second Edition, Technomic Publishing Company, Inc. Lancaster, PA, 1986, p. 95

14. Otis, J. Richard, et al. *On-Site Disposal of Small Wastewater Flows.* Prepared for EPA Technology Transfer, University of Wisconsin, 1977.

15. Laak, R. Personal Communication.

16. Harkin, J. "Performance of Mound Systems." Presentation to the Wisconsin Groundwater Coordinating Council. January 1994.

17. Hoxie, D. and Frick, A. "Subsurface Wastewater Disposal Systems Designed in Maine by the Site Evaluation Method: Life Expectancy, System Design, and Land Use Trends. *Proceedings of the Symposium of Individual and Small Community Disposal Systems, 1984.* American Society of Agricultural Engineers. pp. 362-371.

18. Tyler, E.; Laak, R.; McCoy, E.; Sandhu, S. "The Soil as a Treatment System. Proceedings of the Second National Home Sewage Treatment Symposium, Chicago, IL, December 1977. American Society of Agricultural Engineers. pp. 22-37.

19. Sauer, D. Intermittent Sand Filtration of Septic Tank and Aerobic Unit Effluent Under Field Conditions, Master's Thesis. University of Wisconsin-Madison, 1975, pp. 32-33.

20. Magdoff, F., Keeney, D., Bouma, J, and Ziebell, W. "Columns representing mound-type disposal of septic tank effluent: II. Nutrient transformations and bacterial populations. *J. Environ. Qual., Vol. 3.* pp. 228-234.

21. Willman, B., Peterson, G, and Fritton, D. "Renovation of septic tank effluent in sand-clay mixtures. *J. Environ. Qual., Vol 10.* pp. 439-444.

22. Reneau, R., Hagedorn, C., Degen, M. "Fate and transport of biological and inorganic contaminants from on-site wastewater disposal of domestic wastewater. *J. Environ. Qual., Vol 18,* pp. 135-144.

23. Cogger, C. "Seasonal High Water Tables, Vertical Separation, and System Performance." Puyallup: WSU Puyallup Research and Extension Center.

24. U.S. Environmental Protection Agency. *Onsite Wastewater Treatment and Disposal Systems*, EPA 625/1-80-012. Washington D.C.: Office of Water Program Operations, 1908, pp. 215-216.

25. Otis, R; Converse, J.; Carlile, B.; Witty, J. "Effluent Distribution," ASAE, 1977, pp. 61-85.

26. Otis, R. "Design of Pressure Distribution Networks for Septic Tank-Soil Absorption Systems, #9.6. Madison: University of Wisconsin Small Scale Waste Management Project, 1981.

27. Converse, J and Tyler, E. Wisconsin Mound Soil Absorption System Siting, Design, and Construction Manual. University of Wisconsin Small Scale Management Project. January 1980.

28. Converse and Tyler

29. Salvato, J. Jr. "Experience With Subsurface Sand Filters," *Sewage Ind. Wastes*, **27**, No. 8, pp. 909-916.

30. Allen, T. *Effective Grain Size and Uniformity Coefficients' Role in Performance Intermittent Sand Filters*, New York State Dept. of Environmental Conservation, Albany, N.Y., June 1971.

31. Sauer, D., Boyle, W, and Otis, R. "Intermittent Sand Filtration of Household Wastewater," *ASCE Journal of the Env. Eng. Div.*, August, 1976, pp. 789-803.

32. EPA. *Design Manual: Onsite Wastewater Treatment and Disposal Systems*. EPA 625/1-80-012. Washington DC: Office of Water Program Operations, October 1908, pp. 119-121.

33. Laak, R. *Wastewater Engineering Design for Unsewered Areas,* Second Edition, Lancaster: Technomic Publishing Co., 1986. p.92.

When soils at a particular site are not suitable for a traditional soil absorption system, alternative methods of effluent dispersal can be used. In this application, the effluent from a septic tank enters a Multi-Flo® aerobic treatment unit. After the aerobic treatment, the effluent is sent to an evaporation bed, consisting of small stones. This method is often used on sites where other types of onsite treatment have failed.

CHAPTER 8

DESIGN OF AN ONSITE WASTEWATER TREATMENT SYSTEM

Topics of This Chapter

- Basic Design Considerations
- The Effects of Hydraulic and Material Loading
- Need to Customize a Designs for Specific Sites

While the design of an onsite system may appear simple, variations in the particular application and wastewater characteristics can sometimes make operation complex. To minimize operational problems, the designer must fully characterize the entire soil and site conditions in which the onsite system will be installed and the nature of the hydraulic flow and wastewater loading to which the system will be subjected.

WASTEWATER FLOW

Wastewater flow (or simply "flow") is seldom constant. Water use in a residence will rise when the residents are bathing, cooking, and cleaning, and will drop when they are away or asleep. For a restaurant, water use will rise when customers are there because the customers order meals, which results in dirty dishes, and use of the toilets. For a business, water use rises when employees are present. For an industrial building, water use varies with the process underway.

In a typical residence, water use, hence wastewater flow, follows a pattern shown in Figure 8.1. As this figure shows, most of the wastewater is generated in the morning and evening. Flow during the day and at night is minimal. Likewise, the water use in any establishment can be modeled and estimated if water meters are installed and monitored.

Because reliable data are difficult to obtain for various types of buildings—or because the flow is so variable—many sanitation codes specify the flows that must be assumed for a particular structure. Sometimes these values are quite accurate; sometimes the values are "projected worst case" situations or based on dated information. The only way to accurately estimate flow is to measure it

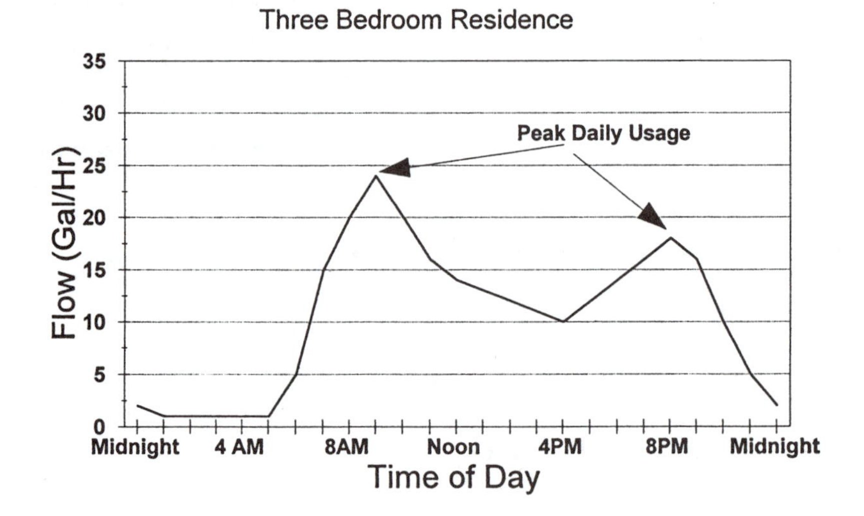

Figure 8.1 - Typical domestic wastewater usage pattern

directly or to compare it to similar structures operating under similar conditions. If published data are used, the designer will have to consider the basis for the flow -- the use of water-wasting plumbing fixtures, for example -- and recalculate the flow if water-conserving fixtures were installed.

There are five types of flow that must be considered in designing an onsite system. Each has a bearing on the design and must be considered, even if judged to be trivial[1]. These types of flow are:

Measured Daily Flow (MDF). The MDF is the actual flow from an structure for an extended period of time such as a year or tourist season. It should be graphed so that the variation in flow is readily apparent.

Average Daily Flow (ADF). The ADF is the *annual flow* divided by the number of days water was used. For a typical residence, this value will be the annual flow divided by 365 days. For seasonal buildings, such as campgrounds or mobile home parks, the value may be the total flow divided by the number of days the facility was open. ADF can be used as a rough guide, but as a statistic, ADF is meaningless by itself.

Maximum Month Average Daily Flow (MMADF). The MMADF is the 30-day period having the highest average daily flow based on an annual flow. For larger onsite systems and municipal designs, the MMADF is the "design flow" for the basic process design. This value represents the highest recurring flow that must be taken into account for the design.

Peak Daily Flow (PDF). The PDF is the highest daily flow that can be expected from a structure. This value represents the maximum flow that must be considered in the design. This is not to say that the system must be designed to provide daily treatment at the PDF; rather, the PDF is the highest flow that the designer will have to account for in the design.

Peak Hourly Flow (PHF). The PHF is the highest hourly flow the onsite system can expect to receive during the life of the system. This value can be significant for large systems where inflow and infiltration (I&I) from storm events can markedly increase flows.

Table 8.1 shows the relationships among the different flow types[2].

Table 8.1 - Flow Rates

Flow Type	Ratio to ADF
MDF	Varies
ADF	1.0
MMADF	1.4
PDF	2.0
PHF	3.0

Consider, for example, a church. For six days a week, the flow may be low, rising whenever there is a special event. On Sunday, when activities are highest, the flow will also be the highest. The designer has several options to consider. The onsite system could be designed to treat the peak daily flow, but this alternative could result in an onsite system that is too large to function properly the remaining six days of the week and be prohibitively expensive. A better design would include an equalization basin that would hold the peak daily flow and discharge it to the treatment units at a controlled rate. An effluent pump must be installed to dose the treatment system so that the holding tank will be emptied during the week. Using this design, the remaining unit processes can be designed based a daily flow that is 1/7 of the peak daily flow.

Sometimes owners of seasonal buildings try to undersize their onsite system to save money. Undersizing may work if there is an adequate resting time between uses, but undersizing poses risks. Consider the water usage in a summer resort, as shown in Figure 8.2. Between Memorial Day and Labor Day, the ADF is 5,000 gal/day. Memorial Day and Labor Day are both holidays when flow is at its highest, 7,500 gal/day. After Labor Day, the ADF drops to 300 gal/day. The owner suggests that the design be based on the average daily flow of [(9/12) * 300] + [(3/12) * 5,000] or 1,475 gal/day. This undersizing would result in system failure. A more prudent approach would allow for a minimum capacity of 5,000 gal/day and provide holding tanks or equalization basins for peak events. Or, the designer could divide the onsite system into smaller units, dosing and resting each to provide a small enough unit to treat the wintertime

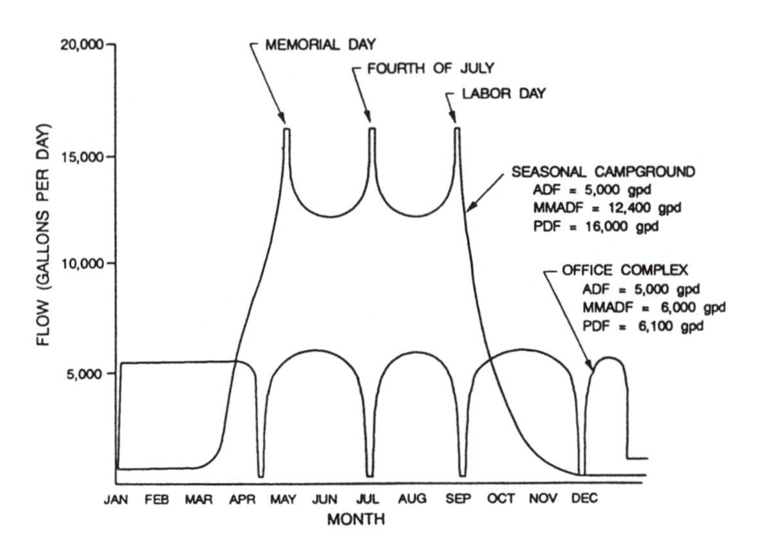

Figure 8.2 - Summer Resort and Office Complex Water Use Patterns

dosing and resting each to provide a small enough unit to treat the wintertime wastewater load, putting other units on line as necessary.

As important as understanding the flow to the onsite system, the designer must also understand the character of the wastewater the onsite system will receive. Most onsite system designs, as we have seen, are based on "typical domestic sewage." Wastewater that contains more organic material, nutrients, metals, or toxic substances than "typical" will damage the treatment ability of the onsite system. The nature of the wastewater must be fully predicted and understood before the system is designed[3,4].

The wastewater characteristics of "typical" installations can be estimated within margins of error, even for commercial or industrial installations. The best method for characterizing the wastewater for a particular place is to obtain enough wastewater samples to be certain of the range in the wastewater's characteristics. These samples must be taken at different times and days, depending on "significant events" at the facility, such the cooking meals, peak customer demand, or industrial processes.

Wastewater Characteristics

Wastewater Characteristics are detailed in Chapter 2, *Water and Wastewater Characteristics*. This section is devoted to assessing the particular challenges that are posed by various applications.

$$Human\ Equivalent = (8.34 * 10^{-3}) * \left(\frac{w}{m}\right) \quad (8.1)$$

size of a project. A typical adult will excrete on a daily basis approximately 77 g (0.17 lb) of BOD, and 91 g (0.20 lb) of total suspended solids[5]. The human equivalent of a given wastewater strength can be calculated thus:

where:

w = constituent concentration, mg/L (BOD_5 or TSS)

m = assumed human concentration, lb/day (.17 for BOD_5 and .2 for TSS)

Food service facilities are most prone to having undersized systems when wastewater strength is not taken into account. Consider a fast-food restaurant that serves hamburgers, french fries, and milk shakes. The sanitary waste — wastewater generated from toilets — is about 1700 L (450 gal)per day and meets the criteria of typical domestic wastewater. The kitchen produces about 380 L (100 gal) of wastewater having a BOD of 1500 mg/L, mainly from oil and grease and total suspended solids level of 2,000 mg/L. The sanitary waste is equivalent to a typical 3-bedroom home, but the kitchen waste is far stronger than household waste and would quickly overwhelm the onsite system unless special precautions are taken. From Equation 8.1, the total BOD loading is the product of $(8.34 * 10^{-3})*(1500/.17)$ and equals 73.59 persons. The suspended solids loading is $(8.34 * 10^{-3})*(2000/.20)$ and equals 83.40 persons, a loading too high for a system designed on the basis of flow.

One precaution is to use an aerobic treatment unit to oxidize the oil and grease so it can be mixed with the sanitary waste. If this approach is unacceptable, the kitchen waste can be discharged into a holding tank that from which The wastewater can then be transported to an offsite treatment facility.

Industrial facilities pose special problems because the nature of the wastewater varies with the type of process. In every case, the nature of the process must be fully understood, and pretreatment using specialized devices before discharge to the onsite system must also be considered.

Garage catch basins pose a particular challenge. Depending on local regulations, the catchbasins either discharge to the onsite system or to a drain tile. Such discharges often contain petroleum and other automotive fluids. Increasingly, owners of such facilities must install holding tanks where the wastewater is stored for offsite treatment[6].

Occasionally, the quality of the source water must be factored into the design of an onsite system since some waters contain excessive concentrations of dissolved minerals, usually calcium and magnesium (*hardness*), or be acidic. Either situation poses challenges for the designer. For example, hard water can leave deposits on pipes, fitting, and appurtenances that can affect the efficiency of the equipment. If the minerals in the water react with wastewater constituents, the resulting products may precipitate out, plugging up orifices or abrading the interior of the system. Hard water is often treated by a water conditioner that substitute sodium for the calcium and magnesium. The backwash waste from a water conditioner, $CaCl_2$ and $MgCl_2$, can affect the operation of the system, depending on the concentration of the chloride[7].

Acidic water can corrode metal pipes and fittings — copper, in particular — and create treatment challenges if the copper concentrations in the wastewater effluent exceed discharge limitations. Acidic water can be neutralized or accommodated by using acid-resistant plastic pipes and fittings.

Water Reduction and Recycling

Many designers are examining water reduction and recycling measures in combination with alternative wastewater treatment procedures. Water reduction makes economic sense for all installations because of the expense to pump and consume water and to treat and dispose of wastewater. These costs can be reduced if water use is reduced. Also, water recycling makes sense for buildings where most of the water is used for nonpotable activities. For example, in places of business, most of the water use is for flushing toilets. While this water should be "clean," it does not have to be drinkable.

Water reduction can be achieved through the use of water-conserving plumbing fixtures. Toilets are now available that require only 1.6 gallons of water per flush. Some styles use pressure assistance, further reducing the water consumption to 0.5 gallons per flush.

Packaged treatment devices, such as the Single-Home FAST system from Bio-Microbics, are available that can treat and recycle wastewater intended for nonpotable needs. Such systems require dual plumbing systems to return the water for nonpotable use. Manufacturers claim such methods can decrease water use by over 90 percent, depending on the building use and, as a result, reduce the soil absorption area required. It is not necessary to use a manufactured treatment system because a properly-sized aerobic treatment unit can accomplish the same end if the effluent is redirected to the structure through a separate plumbing system. One system, the Zenon System (formerly Thetford), is usually installed in industrial facilities. A schematic for that system is shown in Figure 8.3[8]. The Toronto *Healthy House* is shown on pages 113 and 114, which is an integrated water collection, recycling and treatment system.

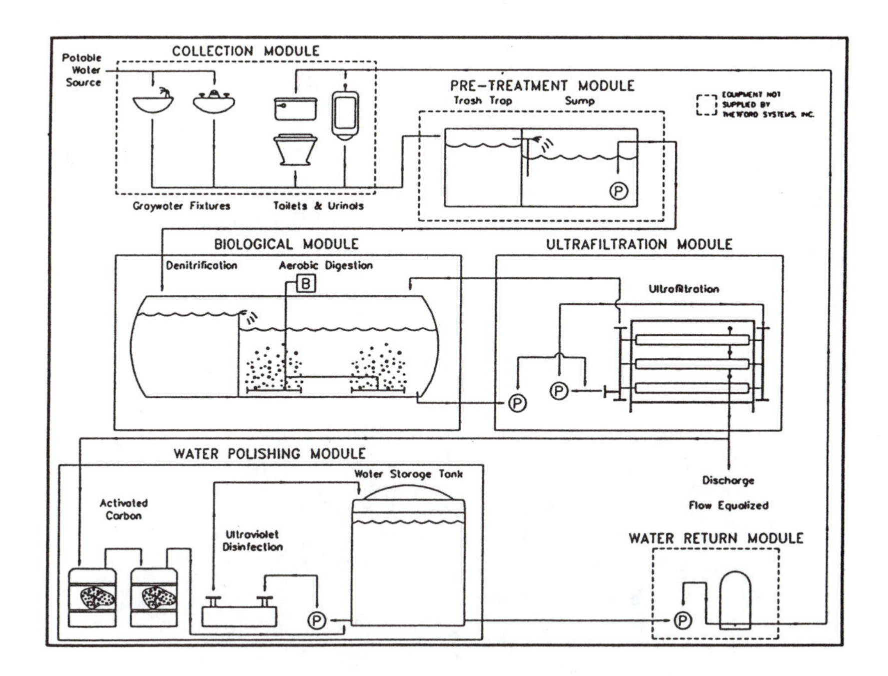

Figure 8.3 - Zenon System (formerly Thetford)
(For further information, call 800-443-3006)

Summary

In designing an onsite system, the designer must examine the flow and the loads to the system. For the flow, the designer must consider the variation in flow and seek to provide a steady flow to the unit processes. In examining the loads, the designers must identify the organic and inorganic constituents the system will be required to treat. Each unit process must be designed to treat the flow and loads the designer determines will be subject to treatment. In conjunction with sizing unit processes, the designer should incorporate water reduction and recycling measures to reduce the water the onsite system will have to treat.

References

1. Qasim, A. *Wastewater Treatment Plants Planning, Design, and Operation.* New York: Holt, Rinehart and Winston, 1985, pp 26-29.

2. Metcalf and Eddy. *Wastewater Engineering Treatment, Disposal, and Reuse.* New York: McGraw-Hill, Inc., 1991, pp 22-29.

3. Metcalf and Eddy, pp 108-110.

4. Qasim, pp 42-44.

5. Salvato, J. *Environmental Engineering and Sanitation*, Third Edition. New York: John Wiley & Sons, 1982, pp 522-523.

6. Shaw, Bryon; J. Tyler, P. Sauer, G. Lueck. "An Evaluation of Current Method Used in Wisconsin and Alternatives for Disposal of Motor Vehicle Waste Fluids." University of Wisconsin-Madison Department of Soil Science, 1993.

7. Metcalf and Eddy, pp 111-112.

8. Zenon (Thetford) Systems, Inc. "Cycle-Let Wastewater Treatment and Recycling Systems." 1987. (800-443-3006)

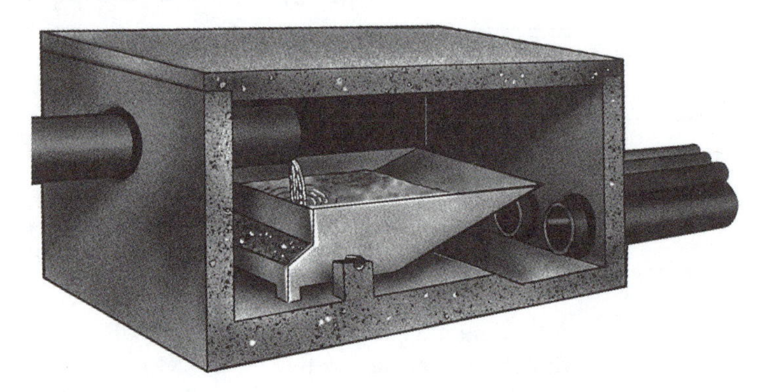

INSURES EQUAL DISTRIBUTION

- Automatically discharges 1.5 gallons of retained effluent from the storage tray because it's designed on the basic pivot and balance principal.
- Makes system work uniformly instead of the trickle effect by providing equal flow distribution to each outlet.

INCREASES LONGEVITY

- Provides indefinite life expectancy for septic systems.
- Laboratory tested to last a minimum of 25 years without failure.
- Maintenance free.
- Minimal Added Cost.

INSTALLATION

- As simple to install as any D-Box.

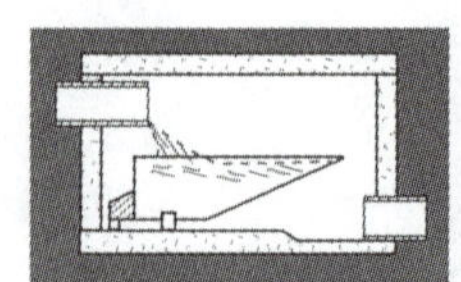

1. The *Dipper* in its upright position filling up at a slow rate with effluent from the Septic Tank.

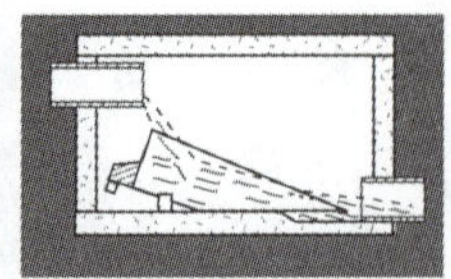

2. When the *Dipper* has retained 1.5 gallons of effluent it will automatically discharge and equally dose the system in 1.5 seconds.

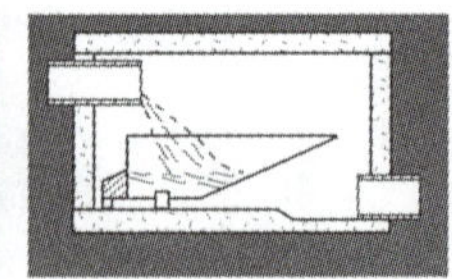

3. Dipper ready to start the next cycle.

Polylok's *Dipper* distribution box helps to provide even dosing to the distribution pipes. Courtesy of United Concrete Products, 173 Church Street, Yalesville, CT 06492

CHAPTER 9

INSTALLATION OF AN ONSITE WASTEWATER TREATMENT SYSTEM

Topics of This Chapter

- *General Installation Techniques*
- *Guidelines for Meetings with Interested Parties*
- *Specific Installation Requirements*

The proper installation of an onsite system is as important as design if the system is to work properly. Each system must be installed in strict conformance with the approved design plans. Reliable installation practices must be followed for the system to operate successfully. The methods, materials and safe installation techniques must be employed.

DESIGN AND PRECONSTRUCTION CONFERENCE

Every successful installation begins with an effective design, so the installer should be involved in the soil and site evaluation process. An experienced installer can provide important insights that will result in a successful design and, ultimately, a properly-functioning system. The design may also have to be reviewed by local or state public health officials before finalization. After the plan is approved but before installation, the designer should conduct a preconstruction conference. At this meeting, the designer, installer, regulator, and owner each has an opportunity to hear each others' perspectives and express their individual expectations for the system. For example, the designer may have critical features the installer must respect; the owner may identify trees or shrubs that the installer must protect; regulators have codes to enforce; the installer has a schedule and requirements for the job.

Figure 9.1 - Preconstruction Conference

The Meeting

The preconstruction conference should follow a written agenda and be held in a location adequate to show the plans to all participants, as shown in Figure 9.1. The designer and installer should each bring the following:

- A Plot Plan of the Site
- A Copy of the Local Code
- A Set of Approved Plans
- Permits and Other Approvals
- A Materials List
- Installation Guides, if any
- Installer's Schedule and Sequence of Installation

The Agenda

A typical agenda will contain the following information and topics:

- Date, Time, and Location of the Conference
- Remarks by the Owner, Designer, Installer, and Regulator
- Critical Components of the Design
- Presentation of Installation Process by the Installer
- Owner Requirements and Expectations
- Code Issues

Meeting Minutes

The installer or designer should maintain and distribute the minutes of the meeting. Key points can be written on a flip chart that can be viewed — and challenged on the spot — by all. If there are no objections, the flip chart notes can form the basis of the minutes and later typewritten minutes will be developed from them, as shown in Figure 9.2.

Meeting Minutes
Date: 5/18/94 Time: 10 AM
Present: M. Stone, Owner
T. Graham, Designer
R. Kaminski, Inspector
J. Quast, Plumber

Topics Covered:

- Installation must be completed on a sunny day.
- Large maple trees on south side of soil absorption area to remain undisturbed.
- Inspection of septic tank, pump chamber, and soil absorption system is required.
- Plumber must provide 24 hour notice before inspection.

Figure 9.2 - Meeting Minutes for Preconstruction Conference

In most cases, the designer and installer are two different people. The designer should insist upon the option, and right, to perform *observation services* during construction. For small installations, these services can be as informal as stopping by the site during construction to check on the installation of critical components. For larger installations, the observation services may include having staff present during most of the construction and startup. In either case, all the parties benefit by ensuring that installation proceeds as intended.

System Layout

Whenever several people are involved in the design process, a high probability exists that information can be overlooked or misidentified. To minimize misunderstandings, before installation proceeds, the site should be staked out to confirm that the plans were mapped accurately. Staking out the location of system components can prevent disputes after the installation begins.

The first step is to flag the system area, particularly the soil absorption area, which is more sensitive to damage than the tank location. The four corners of the septic tank, and any other treatment units, should be staked. Next, trench limits must be identified. If the trenches are along a contour, particular care must be taken to provide enough flags to identify a constant elevation and alignment; trenches along contours usually follow an arc. Trees and shrubs that are to be protected or removed must be properly identified. When the process is completed, the layout should resemble the plan view of the system. If the staked design looks different, the reason must be identified and the plot plan or system layout corrected.

During system layout, errors must be addressed before construction begins. Usually, the installer is a different contractor than the plumber or builder. As a result, information among the parties may not be accurate and details and locations of the onsite system may not be included in the house construction plans. It is common to see sewer laterals that are stubbed on the wrong side of the house and elevations that are different from those planned. Occasionally, the site of the structure may have been moved, sometimes to the place where the onsite system was supposed to go. The onsite system designer may then need to redesign part or all of his original plans. It is always better to change a design before construction than to have a contractor submit a change order and expect payment for "down time" after construction begins.

A layout should identify other potential problems such as stumps, buried utilities, or other items that may have been unknown during the initial site evaluation. Most designers anticipate some hidden obstructions and provide flexibility in their designs to adapt to those obstacles.

INSTALLATION

Before conducting any excavations, the installer must ensure there are no buried utilities in the excavation area. There are services available to locate buried utilities. All contractors know of these free services and are responsible for identifying them.

Levelling

The importance of using a level during installation cannot be overemphasized. The treatment tank, pump, pumping chamber and infiltration system must all be level. Levelling starts with providing the right equipment at the site and experienced people to operate it. Excavations must include sufficient room — at least six inches — on the sides and bottom for granular backfill. Most excavations will be much larger than this. All excavations must be wide enough to prevent sloughing from the sides or spoil pile.

The bottom of the tank excavation must be level. On top of the soil, a 150 mm (6 in) layer of crushed stone should be placed, raked, and tamped to provide a firm, level base for the tank. Gravel should be placed in lifts until the top of the base is at the correct elevation on all four corners and center of the tank location.

Figure 9.3 shows the proper excavation for a treatment tank. There are three reasons why tilted tanks do not function properly. First, the tank will not hold the specified volume if it is tilted. Second, the tank may short circuit because the tilt may provide preferential direction to the flow. Third, sludge tends to accumulate in the lowest elevation, instead of settling uniformly along the bottom.

Figure 9.3 - Excavation for a Septic Tank

When the tank is lowered into the excavation, its bottom should be checked to ensure that it is squarely resting on the stone. The elevations of the four tank corners should be checked and compared to the plan. The tank location should be triangulated, and the top should be checked to see that the tank is plumb.

Once the tank is set correctly, the sides of the excavation can be filled with stone, except for the very top, which is filled with topsoil. Native soil used as backfill should be placed in 150 mm (6 in) lifts, each lift compacted before the next lift of soil is placed.

Treatment and Pump Tanks.

Both treatment and pump tanks must conform to the models the designer has specified. If the tanks are not identical models, they must be functional equivalents or an interchangeable model. This is because the treatment tank volume and shape has been selected to perform a specific function. Mere capacity is no justification for swapping one tank for another. The installer must understand the reasons a particular tank model was selected and respect that choice. These functions include:

- Providing a detention time of a certain duration,
- Having a bottom surface area of a designated configuration,

- An adequate storage capacity, and
- A specific volume per inch of depth.

Pipe Couplings

All treatment tanks have piping to and from the tanks. If the piping is not flexible, it can break when the tank settles. To prevent such breakage, flexible coupling should be used within the first 150 mm (6 in) of the pipe opening. Flexible couplings allow the pipe and tank to shift independently without damaging each other. Since the flow from the structure is by gravity, the opening to the tank must be below the outlet from the structure.

Orientation

Tanks should be oriented as shown on the plans. Orientation is important when space is limited or other site limitations prevent the installation of additional pipes and fittings. Orientation is critical to the proper operation of a gravity-fed system. A misorientation of a treatment tank can affect the performance of the system, and correcting the mistake will require additional pipes and fittings or more openings in tanks at unplanned locations.

Soil Absorption System

The soil absorption system is the most crucial activity of any system installation, whether it is the excavation of a trench or the surface preparation for a mound system. A proficient installer will check the moisture of the soil first by seeing if it can be rolled into a wire. If the soil forms a wire, the soil is too moist. The site can be permanently damaged if the soil is too moist, insuring that the system will fail soon after startup. Moist soil can be smeared, sealing the surface, thus reducing the infiltrative capacity. Whenever the soil shows too much moisture, it should be allowed to dry out. Once the soil no longer forms a wire, the site is ready for installation.

Trench preparation.

Trenches are usually prepared by digging with a backhoe. For unusual sites, such as on a steep contour, trenches should be dug by hand. In either case, the bottom is dug level, and large clods are removed. All activity, like walking in the trench, is kept to a minimum. Before the gravel or leaching chamber is installed, the operator should scour the bottom of the trench with the backhoe's bucket teeth to roughen the surface. Many installers also attach teeth to the sides of the bucket to roughen the sides. After the bottom and sides are prepared, either gravel or leaching chambers are installed. If gravel is used, it must be carefully poured into the trench and raked level. Gravel that is just *dumped* into the trench can compact the bottom surface and seal it against infiltration. After the surface of the gravel is levelled, the distribution pipe is laid atop the gravel and more gravel is placed above it. Geotextile fabric is placed above the gravel to prevent fine soil particles from migrating into the gravel and sealing the pore spaces. Figure 9.4 shows the installation of a typical gravel/pipe distribution system.

Figure 9.4 - Gravel-Pipe distribution method

If a leaching chamber is installed, it is placed into the trench and connected according to the manufacturer's installation instructions and then backfilled with suitable soil. Figure 9.5 shows the installation of a leaching chamber distribution system.

Figure 9.5 - A leaching chamber distribution system (Courtesy of Infiltrator Systems, Inc. Old Saybrook, CT)

Piping

All pipes should be inspected to ensure that they are the proper size and grade, and that they conforms to local plumbing codes. If a pressure distribution system is used, the distribution piping must be inspected for proper orifice size, location, and spacing. All of the orifices should be along a single axis. The holes should be cleanly drilled, free of burrs or irregular shapes and straight. All joints must be clean before they are solvent welded, and solvent must be applied in accordance with the manufacturer's installation instructions.

Stays

Pipe stays should be used if the installer is having any difficulty keeping the distribution pipes in place. The stays can be used to set the elevation and alignment of the pipe as well as holding it in place while the trench is being backfilled.

Cleanouts

Cleanouts are necessary for the maintenance and initial calibration of a pressure distribution system. The designer must include cleanouts in the design and the installer must include them for any pressure distribution pipes. Figure 9.6 is a diagram of convenient places for cleanouts in a pressure distribution system.

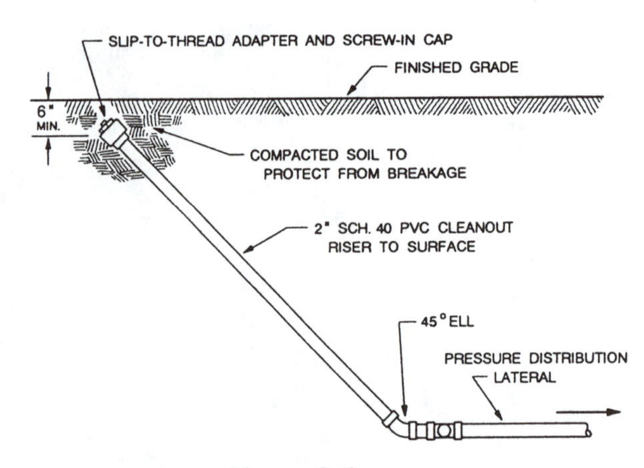

Figure 9.6
Cleanouts for a Pressure Distribution System.

Pumps

Proper pump installation is important for a successful pressure distribution system. The pump must be exactly as specified: make, model, impeller, impeller material, and voltage. The pump must be set at the correct elevation and set to turn on and off at the right intervals. The pump must be equipped with an alarm in the event of a failure, and it must be connected to the proper power source, whether it is 12 volts, 120 volts, or 220 volts. (There is little application for 440 volts in most onsite applications.) If the installer cannot locate the specified model, he must find a model that is a functional equivalent, virtually identical to the specified model.

The pump must be installed at the elevation shown in the design. The proper elevation for the pump is largely a function of the proper elevation of the pump chamber. Changing the elevation will seriously affect the pressure distribution function. For example, a pump that is installed too high will send out more effluent than the designed system can handle. A pump that is set too low will not be able to provide the flow specified in the design.

Pressure distribution systems are designed to provide a certain dose volume. This dose volume is a function of the depth of the wastewater in the pump chamber. The depth of the wastewater is dependent on the elevation of the on and off float switches. It is important that the switches are set correctly when the system is installed. If the system is installed in accordance with the approved plans, the pump switch elevation should be correct. However, final switch setting will confirm the design, not be dictated by it, because the final elevation will vary depending on how closely the actual installation conforms to the plan. The possible changes during construction could affect the final elevations.

Electrical

Electrical work should be performed only by licensed electricians. Electrical components inside tanks must be rated for wet and corrosive environments. Buried components must be suitable for underground installation. All components must be properly grounded and fused. Alarms and motors should be on separate circuits or the motor that trips a fuse will also deactivate the alarm system. The National Electrical Code, which is widely adopted, details requirements for electrical installations. Designers and installers should be familiar with this code because much of their work involves electrical components.

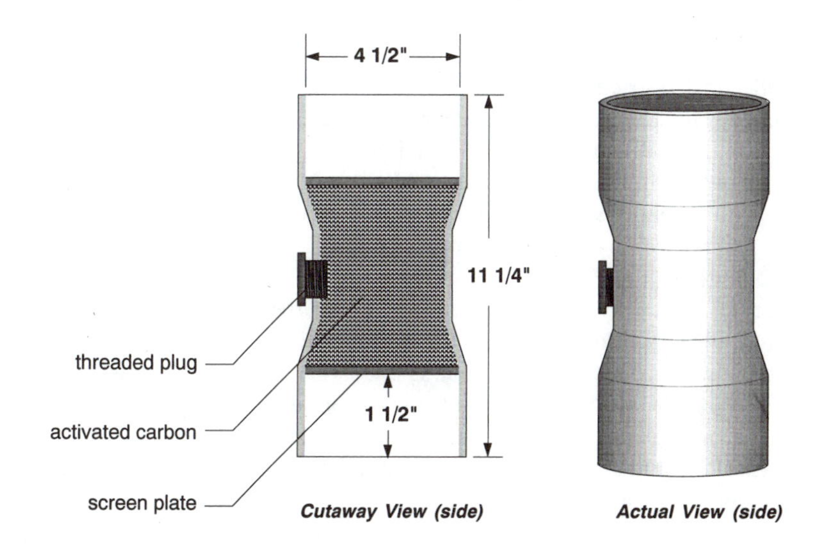

Figure 9.7 - Charcoal filter for roof stack
(Courtesy of Orenco Systems, Inc. , Roseburg, OR)

Venting

An onsite system is designed to vent through the plumbing system in the structure and the soil absorption system. If there is a pump chamber, it will also be vented. Proper venting is essential because an improperly vented onsite system will create an odor nuisance after it is installed. The installer should pay attention to the location and height of roof vents. The vents may need to be extended, moved or provided with an activated charcoal filter, like one shown in Figure 9.7 if the occupants complain of odors.

Local codes may vary but, in general, vents terminate at least eight inches above the roof line. In addition, vents should be at least 10 feet from a fresh air vent or window and five feet from an exhaust vent. Vents should never terminate below an overhang.

Final Inspection and Startup

After the onsite system is installed but before it is put into service, the designer and installer should conduct a final inspection. In many districts, public health officials will also be required to conduct a final inspection. The installation should be compared against the plan, and the plans should be marked to show the actual location and elevation of components. It may be necessary to produce "as built" drawings if many changes were made or the original plans with the construction notes are unreadable. If "as built" drawings are necessary, a copy should be provided to the owner, installer, and concerned regulatory agency.

After the final inspection, the onsite system is ready to be put into operation. The startup should proceed in accordance with the Operation and Maintenance Instructions the designer provided, and the designer should be notified if problems develop. The tank should be filled with water to the outlet level and seeded with several buckets of ripe sludge[1]. The designer should inspect the system several weeks after operation has begun.

Most onsite systems will not require special treatment once they are put into operation. Their performance may vary but should reach stability after several weeks. If an aerobic treatment unit is used, the unit can be "seeded" with several gallons of activated sludge obtained from a local wastewater treatment plant. The activated sludge will multiply quickly and cut the performance time from weeks to days. Sand filters and septic tank-based systems will achieve their performance naturally.

Reference

1. Wagner, Edmund G. and J. N. Lanoix, *Excreta Disposal for Rural Areas and Small Communities*. World Health Organization, Geneva, 1958. p. 137.

CHAPTER 10

OPERATION AND MAINTENANCE

Topics of This Chapter

- *Requirements for Operating and Maintaining an onsite system*
- *The components of an O&M Manual for the system user.*
- *Steps in educating the owner on O&M requirements*

Successful design and installation of an onsite wastewater treatment system are only two of the three essential steps necessary to ensure that the system will function as the designer intended. The third step is proper operation and maintenance of the system.

Operation and Maintenance (O&M) begins in the design phase as the designer identifies and selects the components that will comprise the onsite system. Each component has a design life and an associated maintenance schedule. The designer will select the models based on certain assumptions. These assumptions include the flow and character of the wastewater. The designer must advise the owner of all the assumptions that were made in designing the system and the maintenance schedule for the various components of the system.

Most onsite system designs use native soil as the treatment and dispersal medium. Because soil is not a consistent, manufactured material, it is very important to understand the "design life" and maintenance schedule of the soil. This schedule will depend on how effluent is applied to the soil and if there are any specific local codes that regulate the use of alternate soil absorption systems. The design may also include custom components, like a sand filter, which may be unique to that system. The designer will need to tell the owner how to keep the custom-designed components functioning correctly. An O&M manual should include the following:

- Set of Approved Plans
- List of manufacturer and model for each manufactured component
- Maintenance schedule for each manufactured component
- Maintenance schedule for each custom-designed component

OPERATION AND MAINTENANCE MANUAL

A very basic O&M Manual consisting of four minimum requirements may help the trained technician, but a really effective O&M manual must recognize that the owner is the principal user even if his ability to understand the technical data is limited. To interest an owner in protecting the information about the system, a manual — like the one shown in Figure 10.1 — must explain why the information is important. An O&M manual is an educational tool for the owner; the designer must explain the system and how the components work together. Many regulatory agencies provide brochures on onsite system care and operation, so the designer may not have to create the manual from scratch. While providing one of these documents may be sufficient in many cases, the designer should attach other helpful hints or information specific for that particular system.

Operation and Maintenance Manual

Calico Lane Subdivision
Jermyn, Pennsylvania

Recirculating Sand Filter
Subsurface Infiltration
MMADF: 16,000 gal/day

Revised: January 1994

Designer:
Jones Engineering
Scranton, PA
717/348-1212

Operator:
J. Quast
Jermyn, PA
717/876-1213

Figure 10.1 - O&M Manual

More sophisticated and customized designs may require a custom-written O&M manual. This manual will contain all four elements plus all additional information about the unique aspects of that system. **A copy of the manual for each installed system should be kept in the designer's files.**

Overview

The O&M manual must define the system in terms of its type, size, treatment performance expectation, and location. The overview, which summarizes the onsite system, must have sufficient detail to allow the service provider to trouble shoot problems. But, the manual should be written in plain English and not burden the service provider with useless details.

The overview should explain how the system works, what its intended performance is, and why each component is important to the performance of the system. Each major unit process or component can be described sequentially with a brief explanation of the performance expected from the unit. The expected outcome can be in terms of treatment performance, pumping capacity, application rate, or other major measurement used to design the system.

The overview must also detail the limitations of the design. No system can be expected to accommodate every conceivable combination of flow of constituent loading. The designer must clearly state the design assumptions — or operating limits — of the design. The design might include flow meters and other devices to warn the owner that the system is being operated outside of its limits, but the owner/reader should not rely on alarms alone. An understanding of the limits of the system may obviate the need for the alarms.

The overview should show all the functions of the system and the interrelationships among the components. This way, the first part of the manual should tell the reader why it is so important to respect the requirements and limitations of the design.

Operation begins with the initiation of the system, so the next section must show how to start the devices. Onsite systems that depend on biological processes — which includes all but holding tanks — can be started with "seeding" of bacteria from a municipal treatment system.

Plot Plan and Drawings. A plot plan and drawings of the system show the location of system components. The plot plan and drawings must be accurate. Most of the components, tanks, pipes, and soil absorption systems, will be buried, so accurate plans must show the location and elevation of these components so they can be located to be repaired.

The plans should be "as built" drawings, that is, drawings completed <u>after</u> the installation. *As builts*, as they are called, provide the most accurate record of the installation. Only rarely is a system installed exactly according to the design plan. Hidden obstruction, errors, omissions, and unforeseen changes all result in modifications. Some common changes at sites are: tanks shifted because the structure was moved from its planned location, final elevations altered to accommodate an improperly buried lateral, characteristics of the soil are discovered to change more than anticipated. The original design drawings can be annotated with comments and corrections as the installation proceeds, then augmented by inspection reports completed by local public health officials.

Material and Equipment Specifications. Material and equipment specifications provide the detailed technical requirements for pumps, pipes, coatings, electrical systems, process equipment, and other items that comprise the onsite system. In simpler designs, these specifications can be noted on the plans themselves; in complex systems the specifications will be contained in a separate booklet.

Material specifications vary in their importance; some items can be exchanged easily, for example, brands of PVC pipe. Other items, such as pumps, filter media, or custom-wired electrical panels, would all be specific for a particular design and must be followed exactly. The designer should make clear the rationale behind each equipment selection. This is the best method of ensuring that the design decisions will be respected as the system is maintained.

Material and equipment specifications should also be summarized in a listing that includes **manufacturer, model number, and performance specification.** This summary can be used a quick guide to the components requiring maintenance during the operation of the onsite system.

Maintenance Schedules

Nothing can be more annoying to the owner of an onsite system than a maintenance schedule. Owners accustomed to municipal "flush and forget" sewer systems might want to continue their water use habits. Many inexperienced onsite system users have been led to believe that onsite systems are maintenance-free because, in the case of the traditional septic system, there are no mechanical parts to break. On the contrary, all onsite systems require maintenance, the schedules differing only in frequency depending upon system use, complexity, and need for effluent quality testing.

Manufactured Equipment. Manufactured equipment usually comes with a set of instructions that set up a maintenance schedule. This schedule is often quite simple; most manufacturers try to make equipment "maintenance free," and "maintenance" can be limited to an activity such as checking for loose connections. Despite the phrase, "maintenance free," components should be examined at least semi-annually because of the conditions or abuse they may have been exposed to, including damage caused by rodents and insects, plugged pipes caused by disposed products, animal nests, corrosion to exposed electrical alarm circuits, or burned-out lights.

Corrosive gasses, such as hydrogen sulfide (H_2S) can be a major source of damage in systems that are not properly sealed and vented. The gas can migrate to any space and damage equipment quickly and silently. Because this damage can be subtle, maintenance must include a visual inspection for the signs of corrosion.

In regions subject to significant freeze-and-thaw cycles, frost heaves can damage components, especially watertight tanks. In these areas, inspections of access risers and tank lids should be conducted during the spring thaw when damage can first be observed and corrected.

Likewise, autumn is a good season for checking onsite systems for protection against freezing. If necessary, tanks should be insulated if they are prone to freezing. Insulation can take the form of additional soil or hay. Snow is an excellent insulator provided it falls before the soil starts to freeze.

Custom-Designed Components

Just like manufactured components, custom-designed components need to be maintained periodically. Unlike the manufactured components, there may be no experience in establishing what this cycle should be, so it is up to the designer to establish one during the first year of operation. If the system is checked once before and after winter, it is on a semi-annual schedule by default. For most components this should be adequate for cursory inspections. A more-detailed inspection should be conducted at wider intervals.

For example, the State of Wisconsin mandates that septic tanks be pumped every three years[1]. Figure 10.2 shows an operator preparing to pump out the contents of a septic tank. The purpose of this pumping is twofold: remove the accumulation of solids that may be impeding the performance of the septic tank and provide an empty tank for the inspection of the baffles, joints, and the general condition of the materials of construction. The interior components are normally submerged and not visible during the semi-annual inspection.

Figure 10.2 - Pumping out the septic tank

Example Operation and Maintenance Schedule

A sample operation and maintenance schedule is:

Every Six Months:

Inspections:

Trash traps or grease traps
Interiors of septic tanks and pump chambers
Filter media of aerobic treatment units and sand filters
Soil absorption system vents for evidence of ponding
Pressure distribution system cleanouts

Tests:

Alarms
Warning Lights
Float switches
Pumps
Valves
Distribution Systems

EFFLUENT QUALITY TESTING

With the emerging emphasis on the environmental impacts of onsite systems, effluent quality testing is becoming a part of the operation and maintenance requirements for larger onsite systems. Onsite systems that have a surface discharge can be tested much like their municipal counterparts. Onsite systems incorporating a soil infiltration system pose many difficulties in obtaining accurate samples; therefore, regulatory agencies may impose limits on the quality of the effluent *discharged* to the soil and make informed guesses about the extent of additional treatment in the soil.

When effluent quality testing is required, it usually centers on the following constituents, which are the traditional parameters for measuring treatment performance:

- Biochemical Oxygen Demand
- Total Suspended Solids
- Total Bacteria

Nutrient removal is now a major concern for two reasons. First, nutrients such as phosphorous and nitrogen can cause eutrophication of lakes, streams, and estuaries. Such problems are discussed in Chapter 2, *Water and Wastewater Chemistry*. Second, nitrogen, in the form of NO_3^{-1}, may pose a health risk to infants at concentrations exceeding 10 mg/L-N. Therefore these nutrients may also be monitored.

If Monitoring is required, it should be conducted in accordance with a *monitoring protocol* that sets forth the conditions and limitations for testing. The protocol should identify:

- Sample Collector
- Monitored Constituents
- Monitoring Schedule
- Sampling Procedures
- Sampling Points
- Analysis Procedures
- References
- Performance Requirements

Sample Collector. Sampling should be conducted by professionals. Untrained personnel might lead to incorrect sampling and incorrect results for the analyses, providing misleading information for system adjustment. The collector must be thoroughly trained in all aspects of sampling protocol and follow those directions precisely or document any deviation from the usual procedure.

Wastewater Sample

Lab

Number:______________
Name:______________
Address:______________

Station:______________
Date:_______ **Time:**_______

Analyses

Test: By:______________

TDS_______ BOD_______
TSS_______ COD_______
Oil & Grease_______ NO_3^{-1}_______
NO_2^{-1}_______ PO_4^{-3}_______
Cl^-_______

Pb_______ Zn_______
Cu_______ Fe_______

Bacteria______________

Figure 10.3 - Sampling Form

Monitored Constituents. Both the sample collector and laboratory must have a clear understanding of the constituents that are being monitored. Having this understanding will help the collector make sure the proper number and type of samples are being collected. This understanding will stop the testing laboratory from conducting unnecessary and expensive tests.

For a typical residence or business, the analysis will be limited to basic tests. For food service establishments or industrial facilities, the number of test parameters will be larger, similar to the parameters shown in Figure 10.3. Food service establishments are notorious for the concentrations of BOD, principally in the form of oil and grease. Special analysis will be necessary to quantify this concentration, and additional treatment may be necessary if the oil and grease exceed the design limitations. Industrial facilities may discharge metals, toxins, and volative organic compounds (VOCs).

Monitoring Schedule. A monitoring schedule can be the most perplexing aspect to identify. Regulatory agencies try to simplify the process by specifying sampling periods based on guesswork. The period may be monthly, quarterly, semi-annually, or annually. Regulatory agencies prefer shorter sampling intervals so malfunctioning systems can be identified sooner; owners prefer longer intervals so they will have lower costs.

The central question is: *How does one know the onsite system is operating in accordance with the design specifications?* Answering the question requires the use of a run chart, which is a plot of of effluent quality over time[2]. The key to successfully using a control chart is understanding the concept of *stability*. If the onsite system is operating predictably, it is said to be stable. If it is acting unpredictably, it is said to be unstable. Seven consecutive events of predictable results generally constitute a stable system, and if the stability meets the design specifications, the system is successfully operating.

If stability is the aim, the monitoring frequency may fluctuate depending upon the unpredictability of the system. Figure 10.4 demonstrates this concept. As the figure shows, a new onsite system may behave erratically at startup. There may be operational problems that have to be addressed and adjustments to be made. Biological colonies must form, and biological and hydraulic equilibrium must be established. Because the unit is new, it must be closely monitored for signs of malfunction. In this case, the monitoring period is every other day.

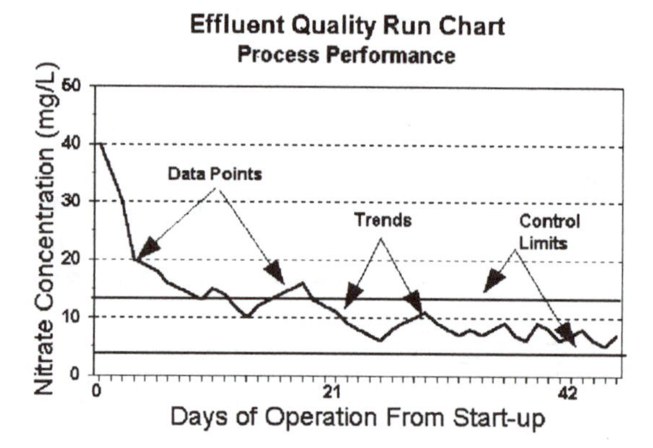

Figure 10.4 - Effluent quality run chart

As the onsite system matures, the performance starts to become predictable and moves towards the stability goal. As this occurs, the monitoring interval is increased to twice a week. Eventually, the system is operating successfully, and the monitoring period is reduced to once a week. As the performance continues to be demonstrated, the monitoring schedule is reduced to once a month, then once every three months.

For manufactured onsite systems, such as aerobic treatment units, it is possible to apply statistical parameters to shortcut the process. This entails demonstrating that, as a whole, all of the manufactured systems operate within a specified range of performances if the influent flow and characteristics are held within specified limits. A detailed discussion of the topic is outside the scope of this text, but can be found in Reference 1.

Sampling Procedures. Sampling can take several forms, each of which has its strengths and weaknesses. Sampling can be an automated process, or it may be done manually. In either case, the analyzed sample may be the result of a single sampling event — a *grab sample* — or the sample may actually be a composite of several sampling events. Also, sampling can be based on time, a predetermined time interval or on a set flow.

A *grab sample* can be used to characterize the quality of the effluent at the time the sample was taken. Such a process works well if the system is stable. If the system is not stable, a composite of samples taken from two or more intervals may be more accurate, because a composite sample will reflect the quality of the effluent for a specific time period.

The *composite sample* can be generated by two methods. If the flow is almost equal during the sample events, the composite can be developed with equal portions from each event. If the flow varies, and this variance is known, the composite can be developed proportionally, based on the flow during each sampling event. Proportional composites are the most accurate for establishing the character of the wastewater when the flow varies.

Manual sampling has distinct advantages. First, the collector can examine the sample and conditions of the system as the sample is being drawn. Second, the collector may be able to conduct a screening analysis, if one is necessary. Manual collection is quite labor intensive. Automated sampling is more efficient for composite sampling. The sampling devices can be located at many places and can take samples as often as required. Continuous monitoring may be necessary to discover the source of "spikes" in the data.

Table 10.1
Sample Preservation Practices[3]

Parameter	Preservative	Maximum Holding Period
BOD	Refrigeration at 4°C	6 hours
COD	2 mL/L H_2SO_4	7 days
Cyanide (CN^-)	NaOH to pH > 12	24 hours
Metals, total	5 mL/L HNO_3	6 months
Metals, dissolved	(pH<2 with HNO_3) filtrate	6 months
Ammonia nitrogen (NH_4^{+1})	40 mg/L $HgCl_2$ at 4°C	7 days
Oil and Grease	2 mL/L H_2SO_4 at 4°C	24 days
pH	None	Determined on the site
Solids	None	7 days
Sulfate	Refrigeration at 4°C	7 days
Sulfide	2 mL/L Zn acetate	7 days

Once the sample is collected, it must be preserved and delivered to the testing laboratory as soon as possible. A sample that is not quickly analyzed may lose its validity. Table 10.1 lists the recommended preservatives and holding periods recommended by the Water Environment Federation. Additional preservation techniques are listed in Appendix B.

Every sample should be clearly marked with the following:

- Owner's name and address
- Date and time of sample, or composite sample length
- Sampling point
- Name of sample collector
- Analyses to be performed

Sampling Points. The location of the sampling point can affect both the quality of the sample and analysis of the results. Sampling points should be clearly identified, preferably on the unit (or units) and in the instructions to the sample collector. All sampling points should be easily accessible, sufficiently large to take valid samples, and easy to service. If the sampling point is difficult to reach, does not intersect the main stream of the flow, or is contaminated, it will produce inaccurate data. A misidentified sampling point will waste the time of the sample collector, result in excessive costs for lab analyses, and cause erroneous analyses of the data that is collected. Each is designed to meet the criteria listed above.

Sampling effluent in unsaturated soil is highly problematic and should not be attempted except by experienced researchers. It is easily possible to affect the quality of the sample by simple errors during sample collection. Accurate sample collection requires the use of sensitive instruments. If unsaturated soil must be sampled, the owner should consult with soil science and engineering professionals who have experience with such testing.

Effluent sampling in saturated soil should also be left to experienced personnel. Sampling wells will have to be drilled, and those activities are regulated and licensed in many states. In addition, saturated soil may be classified as "groundwater" and subject to groundwater protection laws. Anyone contemplating saturated soil or groundwater sampling should contact local or state regulatory agencies for guidance.

Analysis Procedures. All chemical analyses must be conducted in accordance with accepted laboratory procedures. For the analysis of wastewater, the techniques for these analyses are found in *Standard Methods for the Examination of Water and Wastewater* (Standard Methods) and the EPA manual *Methods for Chemical Analyses of Water and Wastes*. These publication present the accepted methods for conducting wastewater analyses and reporting the results of these analyses. The use of standardized methods is critical if the data are to be accurately evaluated and compared.

Performance Requirements. The owner must be provided with a clear understanding of acceptable performance. He must know when the system is considered out of compliance with applicable performance requirements. Does,

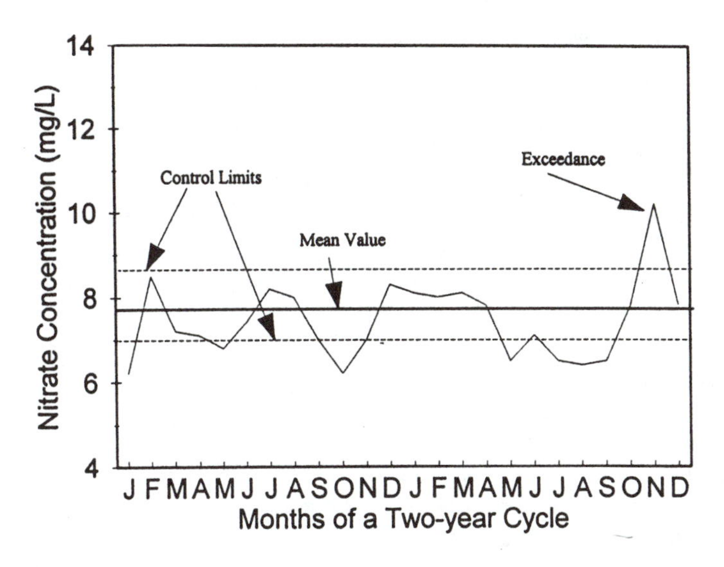

Figure 10.5 - Performance of a typical system

for example, a single exceedance event constitute a failure, or must several events occur? Restated, the question is, "Did an exceedance occur as a matter of chance, or was the exceedance a result of a system overload or failure?" The answer lies in statistics, as can be shown on a run chart. If the system is stable, all data will tend to congregate around a mean value. If the values start to creep in a direction above or below that mean, then a trend is indicated and should be investigated. If a value exceeds the performance standard, the specific value must be compared to all the other values to see if it could be attributed to a specific event. The methods for conducting these statistical analyses are beyond the scope of this text.

Figure 10.5 shows performance for a typical system. As the figure shows, performance varies around a mean value. This mean value, and the variance to the mean can both be calculated and identified as the upper and lower *control limits*. These control limits are values above and below which most operational values must lie. Values falling beyond the control limits are assumed to be the result of failures, overloads, or other events not specified for the design.

Because data is supposed to center around a mean, half of the data will be above it and half of the data will be below it. When data tends to collect on either side of the mean - say seven consecutive points - a new trend is indicated.

References. As indicated previously, sample collection and analysis must be conducted in accordance with standard protocols. Some of these protocols can be specific. It is wise to provide access to reference materials, such as *Standard Methods*, so sample collection and analysis staff can refer to them as they conduct their duties.

EXAMINING EXISTING ONSITE WASTEWATER TREATMENT SYSTEMS

An onsite system designer or installer is often asked to inspect an existing system. The request may be a part of the transfer of property, a sanitary inspection, or from owner concern. This can be exasperating because most, if not all, of the system will be buried, and probably few, if any, records exist. In some cases, the person examining the system will be asked to estimate the condition because there will be no way to examine it, short of digging up a yard.

Existing System Inspection

Name:__________
Address:__________
Property ID No.:__________

Septic Tank:

Capacity:____gal **Age:**____yrs

Pump Chamber:

Capacity:____gal **Pump Size:**____hp

Soil Absorption System:

Trench / Bed / Dry Well (circle)

Length:____ft **Width:**____ft **Depth:**____ft

Comments:

____
____
____

Note: Attach Copy of Soil Evaluation

Figure 10.6 - Existing System Inspection Form

There are ways to minimize the guesswork involved in the evaluation, starting with a check of records the owner has on file or with state or local public health agencies. Owners seldom keep such records, but many agencies maintain copies of percolation test results, soil evaluations, system designs, and inspection reports. It may be possible, depending on the locality and age of the system, to find an accurate plan and inspection report. Failing that, plans, evaluations, and test results might be obtained from the professionals who conducted them. There may be a fee for these reports, or these professionals may be unwilling to provide them, citing a need to verify that the information originally provided was used as intended.

If plans are not found, the owner may have to hire someone to go through the potentially-laborious exercise of locating and inspecting each system component. If so, the owner should select a professional who specializes in the design and installation of onsite systems or a licensed inspector of onsite systems. Owners should be wary of anyone who professes to conduct such inspections but has no professional credentials to back up the claim.

The results of the inspection should be included on a report form that specifies information about the condition and performance of the onsite system[4]. The report should include a sketch showing the location and elevation of system components. If the onsite system uses subsurface infiltration, the owner may want to include in the inspection the results of any additional soil testing or evaluation that was conducted.

The report should pay specific attention to whether the onsite system complied with all the public health codes in effect at the time the system was installed and the status of the system with respect to the current codes. Some jurisdictions "grandfather" existing onsite systems if they were constructed in accordance with codes in effect at the time of installation. Other jurisdictions require that onsite systems be brought up to current codes regardless of code compliance at the time of installation.

Specific information included in the report includes the following:

- Name and address of the owner
- Parcel number of the lot
- Plot plan of the house and the onsite system
- Description of system including location, size and age
- Evidence and extent of previous maintenance
- Size and condition of the septic or holding tank
  - Capacity
  - Baffles
  - Risers
  - Materials of construction
  - Watertightness
- Size and condition of any pump chambers
  - Capacity
  - Electrical panels and connections
  - Size, condition, and age of the pumps
  - Condition of float switches
  - Condition of pipes and fittings
  - Materials of construction
  - Watertightness
  - Alarms
- Location and condition of the soil absorption system
  - Location and depths of laterals
  - Evidence of ponding or surface eruption of effluent
  - Condition of distribution boxes
  - Soil properties, such as percolation rate or soil characteristics
  - Materials of construction of distribution piping
  - Condition of biological mat
  - Evidence of high groundwater or bedrock

- Condition of a distribution piping or pressurized system
  - Residual pressure at the distal end of the laterals
  - Materials of construction
  - Evidence of plugged orifices
- Condition of mound, if one is used
  - Leakage at the toe of the mound
  - Sand gradation
  - Evidence of plugging at the soil interface
  - Mound dimensions

Aerobic treatment units provide additional inspection complications because they often use additional components and have effluent quality specifications set by the manufacturers or designers.

- Condition of sand filters
  - Condition of sand, particularly the sand surface
  - Watertightness
  - Evidence of plants growing in the filter
- Condition of manufactured treatment devices
  - Visual effluent quality (turbidity)
  - Condition of pumps, motors, piping, and fittings
  - Chemical and biological analysis of effluent
  - Watertightness
  - Materials of construction

The evaluator should also evaluate the water habits of the occupants including:

- Flows, if known, including: MDF, ADF, MMADF, PDF, PHF
- Constituent loading
  - Cooking habits
  - Garbage disposal
  - Clothes washing habits
  - Hobbies, such as film developing
- Number and age of occupants

The most difficult part of the evaluation, even if everything is known, is estimating the remaining life of the onsite system. Such estimations are particularly difficult when the inspection involves the transfer of property. The new owners may place more burdens on the system than previous owners. As a consequence, an onsite system that functions for one family may fail with another. For this reason, many evaluators hedge their estimations with caveats.

The best way to estimate the remaining life of an onsite system is to use and analyze the information that has been collected, the more data, the better the estimate. A grading system that assigns points to the information available can be used. As information on the condition of the system decreases and the age of the system increases, a lower grade would be assigned. Conversely, as the age of the system decreases and the amount of information increases, a higher grade would be assigned. The grade would still be a guess of the chances that the onsite system will function until the end of its design life, usually 20 years.

System Evaluation Scorecard		
Feature	**Condition**	**Points**
Capacity		
Baffles		
Risers		
Materials of Construction		
Watertightness		
Pump Chamber Capacity		
Electrical Panels		
Connections		
Pump Condition		
Pump Age		
Condition of Float Switches		

Figure 10.6 - Evaluation of a System

REHABILITATING A FAILING ONSITE SYSTEM

Even if the onsite system receives a failing grade, it still may be possible to rehabilitate the system, depending upon the nature and extent of the failure. The rehabilitation might be accomplished by adding additional treatment units, resting soil absorption areas, adding trenches, changing water use patterns, or by a combination of these measures.

Many onsite systems fail because of hydraulic overloading resulting from continuous use. If the system has been in use for many years, the soil may no longer be able to accept the wastewater discharged to it. If the soil is not clogged with solids, oil, or grease, it is possible to rehabilitate the soil absorption area by starving the bacteria that are clogging the soil. Starvation can be accomplished by two methods. If sufficient room is available, another soil absorption area can be constructed and used. While the alternate area is used, the original area would be allowed to "rest" during which time the bacteria would die off and form humus.

If insufficient area is available, it may be possible to starve the bacteria while the soil absorption area is still in use[a]. To do so, an aerobic treatment unit would have to be installed in conjunction with or instead of the septic tank that was being used. The aerobic treatment unit would remove the organic material before the effluent was discharged to the soil absorption area. The bacteria clogging the soil would still be starved but also saturated with oxygenated effluent, and this would promote endogenous respiration. The end result would be the death of the bacteria and formation of humus[5].

If ponding is the main concern, an additional soil absorption area can be installed. A distribution box or other valving system would be used to distribute the flow to the new area or to alternate flow between the new and the existing soil absorption areas.

If the failure is the result of additional flow, the flow can be altered or reduced, depending on its nature. For example, if the system is being overloaded periodically, a surge chamber and dosing pump can be installed to eliminate the surge into the septic tank and soil absorption area. Alternatively, or in conjunction with a surge tank, water-conserving plumbing fixtures can be installed to reduce total flow to the system. Water conserving fixtures include lavatories, toilets, shower heads, dish washers and washing machines.

Problems with pressure distribution systems can be addressed through replacement of pumps and installation of valves and standpipes to equalize flow.

Mounds and sand filters can also be rehabilitated, although at greater expense. When these systems fail, it is often the result of using a sand that is too fine. If such is the case, the sand can be removed and replaced with sand that meets the ASTM C-33 standard[6]. Such sand has been demonstrated to function well. If the problem with a mound is a leaking toe, the toe of the mound can be extended to enlarge the basal area. If expanding the toe does not solve the

---

[a] This assertion is based partly on the theory of endogenous respiration and partly on anectodal (at this point) observations. Wisconsin is allowing aerobic treatment units to discharge to "failed" soil absorbtion systems. The failure was the classified as an inability to function hydraulicly. All appear to be functioning hydraulicly after the aerobic treatment units were installed. Reference [3] provides additional information on biomat development.

problem, perforated piping can be installed along the perimeter of the basal area to increase the surface area available to the effluent.

Manufactured equipment and treatment units should be serviced as specified by the manufacturers. If the equipment is past its design life, new equipment should be considered.

SUMMARY

All onsite wastewater treatment systems require maintenance. The level of maintenance will vary with the complexity of the system and water use habits of the users. An operation and maintenance manual should be developed for every onsite system, and the system should be operated and maintained as stipulated in the manual.

Systems designed to achieve a specific treatment performance should be tested in accordance with a sampling protocol. The protocol should specify the constituents to be examined and the conditions for conducting the examination. The protocol must be followed precisely to produce valid results.

Many times, existing onsite systems must be evaluated. It is essential that complete records be maintained that document the design, location, and operation of the onsite system. The less that is known about the location or condition of an onsite system, the greater the likelihood that the system will have to be replaced. If the evaluation concludes that the system is failing, it may be possible to rehabilitate the system without having to replace it. The rehabilitation must address the reason for the failure.

References

1. Wisconsin Administrative Code, Chapter ILHR 87.

2. Ishikawa, K. *Guide to Quality Control*. White Plains: Asian Productivity Organization, 1982, pp 61-84.

3. Water Environment Federation. *Operation of Municipal Wastewater Treatment Plants, Manual of Practice*, MOP 11, Washington D.C., 1990, pp 434-437.

4. Jansky, L. "Inspecting Existing Private Sewage Systems." *Wisconsin Plumbing Codes Report*, April 1994.

5. Laak, R. *Wastewater Engineering Design for Unsewered Areas*. Lancaster: Technomic Publishing Co, Inc., 1986, pp 86-95.

6. Converse, J. and Tyler, J. *Wisconsin Mound Soil Absorption System Siting, Design and Construction Manual*, Publication 15.22. Madison: University of Wisconsin Small Scale Waste Management Project, 1990, pp 8-10.

When an aerobic treatment unit is functioning properly, the wastewater in the center of the reactor will be treated and clarified, as evidenced by the clear water that fills the outer ring.

CHAPTER 11

ADVANCED WASTEWATER TREATMENT TECHNIQUES

Topics of This Chapter:

- *Groundwater Modelling Objectives*
- *Nutrient Removal Unit Processes*

ENVIRONMENTAL OBJECTIVES

The effluent from an onsite wastewater treatment system can sometimes have a negative impact upon the environment. The release of nitrogen and phosphorous to groundwater and surface water has become a concern. Both of these elements can cause water quality problems, in both surface and ground waters, since nitrogen and phosphorus dissolved in water are referred to as *nutrients*. This is because plants require both elements for growth. Nitrogen, in the form of nitrate, has also been thought to be responsible for health problems, particularly in infants and the elderly. The presence of these elements in potable and non-potable waters is receiving attention and methods to reduce the levels of these elements from onsite system are becoming more available.

There are two general approaches in reducing nutrient levels in effluents from onsite systems: dilution and treatment. Each approach has advantages and disadvantages. Dilution is easier to accomplish, but pollution still occurs. Treatment prevents the discharge of the nutrient but can be difficult to design and manage. Of the two approaches, most current efforts focus on treatment. Whether by dilution or treatment, the goal is to reduce the nutrient concentration to levels that will present no environmental or health problems.

NUTRIENT REMOVAL

Traditional onsite wastewater treatment systems are effective in reducing organic and solids concentrations and in disinfecting the effluent. Traditional systems never previously had nitrogen or phosphorous removal as objectives. If nutrient removal is one of the main goals of an onsite system design, it can be accomplished. This area of wastewater treatment is developing rapidly.

Phosphorous is released in effluent as the negatively-charged orthophosphate ion, (PO_4^{-3}), which quickly binds to the soil particles. That is because phosphate combines with common positive ions in the local soil and rocks in the area to make insoluble compounds. Nitrogen, on the other hand, when oxidized to the negatively-charged nitrate ion,(NO_3^{-1}), does not combine with positive ions in the soil, so passes through the soil, eventually moving into surface or ground water. Therefore, nitrogen removal must be accomplished by the system before the effluent goes to the treatment medium.

Microbiology of Wastewater and Soil.

In wastewater, the principal source of nitrogen is from the human waste products of protein metabolism, mostly organic nitrogen and urea. An average person excretes about 16 g/day (0.55 oz/day) of nitrogen. Since the average person uses 189 L/day (50 gal/day) of water, this equates to a daily nitrogen concentration of 86 mg/L per person. In the onsite system, this organic nitrogen is transformed to other forms of nitrogen[1].

Nitrogen from wastewater is of environmental concern for two reasons. First, nitrogen is an essential nutrient for plant growth. In water bodies, particularly coastal estuaries, nitrogen is often the *limiting factor* for the growth of aquatic plants, such as algae. (A limiting factor is that substance needed for plant growth that is in *shortest* supply.) When excess nitrogen enters such a system, algae and plants grow quickly, then die and use up a lot of the water's oxygen to decompose, thus depriving fish of needed oxygen.

Second, nitrogen sometimes has been identified as a public health hazard. Sometimes infants under the age of six months, under certain conditions, might be susceptible to an illness call methemoglobinemia, and excess nitrogen has been linked to cancer in older adults. In 1949, the United States Public Health Service published a study associating nitrate concentration in drinking water with infant deaths from methemoglobinemia. This condition can occur because neonates have an immature intestinal tract that is unable to process nitrate as well as older babies. Bacteria in their intestinal tract can convert nitrate (NO_3^{-1}) to nitrite (NO_2^{-1}). In the bloodstream, nitrite can enter the red blood cells and combine with hemoglobin to form methemoglobin, which cannot carry oxygen. As a result, the infant suffocates from a lack of oxygen. The lack of oxygen gives the affected infants an ashen color, hence the name *blue baby syndrome*. Expectant mothers might be advised to avoid foods or water high in nitrates since there is some concern that nitrate can cross the placenta[2].

In older adults, nitrates in the intestinal tract can combine with various amines to form nitrosamines, many of which are known to be carcinogenic[3]. Usually, foods preserved with nitrates have received the most attention as cancer-causing, but nitrates in any form may become a political issue. In the future, public pressure to remove nitrogen-containing substances from onsite wastewater effluent may heavily influence treatment options.

Nitrogen Removal

Nitrogen removal can occur to some extent in a traditional onsite system. Just reducing the BOD from a 120 mg/L influent level to an effluent level of 30 mg/L will result in a 8-10% nitrogen removal. This nitrogen is assimilated into bacterial cells that stay in the septic tank. If the bacterial processes stop, most of the bacteria—that not bound by undigestible cell hulls—would be re-released back into the environment. Traditional onsite systems can achieve nitrogen removal of 20 percent; mound systems can achieve nitrogen removals of from 25 percent to 50 percent. To achieve higher levels of nitrogen removal, a specific removal process must be designed into the onsite system. This process can be biological or physical/chemical[4].

Biological Nitrogen Removal. Denitrification is a two-step process. The first step of the process is to oxidize all of the nitrogen to nitrate ion (NO_3^{-1}). Then, in a process called *denitrification* the second step reduces the nitrate to nitrogen gas (N_2). The specific microbiology of this cycle is discussed in Chapter 3, *Microbiology of Wastewater and Soil.* Nitrification consumes alkalinity, each gram of ammonia converted requiring about 7.14 g of alkalinity (measured as $CaCO_3$). The removal of alkalinity can affect the operation of the system because reducing the pH of the system—making it more acidic—can inhibit bacterial growth. There is a need to carefully monitor N-reducing systems.

Aerobic Treatment. Most of the nitrogen in wastewater is in the form of organic nitrogen. The first stage in recirculation is aerobic treatment to nitrify all of the organic nitrogen to NO_3^{-1}. The aerobic treatment unit must be sized to treat both the organic **and** nitrogen loadings generated in the structure.

For onsite systems, sand filtration is a good choice for aerobic treatment. Sand filters can tolerate inconsistent flow patterns and can be adjusted to maximize the treatment efficiency of the unit. Nitrogenous BOD demands can be met by recirculating the effluent back through the filter.

Anoxic Treatment. After aerobic treatment to fully oxidize the nitrogen, nitrate must be converted to nitrogen gas (N_2), which can then vented from the system. Unlike nitrification, in which the goal is to oxidize nitrogen, denitrification involves the chemical *reduction* of nitrate. Facultative bacteria — bacteria that can use either molecular oxygen (O_2) or an oxygen-containing substance (NO_3^{-1})nitrate)— perform the denitrification. Given a choice, facultative bacteria will choose oxygen over an oxygen-containing substance. The designer must provide a process that creates an environment in which there is no free oxygen but sufficient oxygen substrate. This type of environment is *anoxic* and essential for denitrification. In the absence of oxygen or an oxygen substrate (anaerobic conditions) facultative bacteria die.

Biological nitrogen removal proceeds, then, by subjecting the wastewater to alternating periods of exposure to oxygen and deprivation of oxygen so that nitrification and denitrification can proceed.

One of the simplest systems for onsite systems is the recirculating upflow anaerobic sand filter as shown in Figure 11.1. At first, this system operates like a conventional septic system. Influent enters the septic tank where it is partially treated and clarified. From there the clarified effluent goes to an *upflow* "anaerobic" (technically, anoxic) filter. From the upflow filter, the effluent is pumped into a sand filter for aerobic treatment.

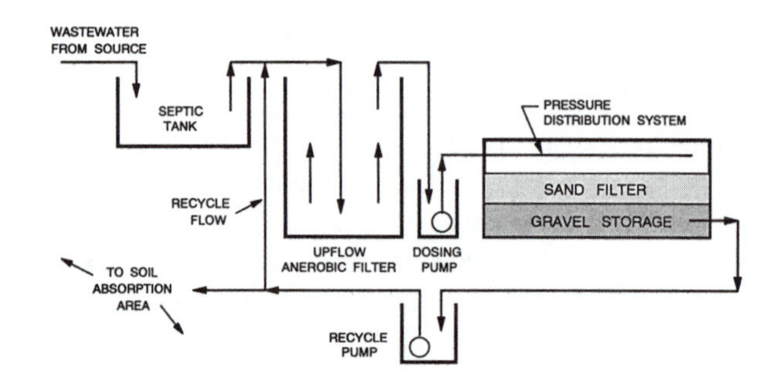

Figure 11.1 - Upflow Anaerobic Filter

As seen in Figure 11.1, effluent from the sand filter is split, some sent to the soil absorption system and some recirculated back to the upflow filter. The bacteria that grow on the medium (usually gravel) are continuously submerged, which protects the bacteria from desiccation during low flow periods. Because the recycled effluent is mixed with septic tank effluent, which typically has a BOD of 140 mg/L, there is enough carbon to support microbial life. Oxygen in the recycled effluent is consumed by BOD in the septic tank effluent, and nitrate, which is also in the sand filter effluent, is consumed when the free oxygen is used up[5].

Because there is molecular oxygen in the sand filter effluent, the recycle ratio - the flow of sand filter effluent to septic tank effluent - must be monitored. If the recycle ratio is too low, the system will insufficiently denitrify the mixed effluents. If the recycle ratio is too high, excess oxygen will be pumped into the upflow filter, stopping all denitrification. Ideally, the recycle ratio should be between 3:1 and 2:1. The system has a theoretical efficiency of 75 percent with a recycle ratio of 3:1. The efficiency cannot be higher because 25 percent of the effluent from the sand filter will not have been denitrified. If 20 percent of the total nitrogen is assimilated by the bacteria and 75 percent of the remaining nitrogen is removed, then the upflow anaerobic filter has a total efficiency of 80 percent.

Other nitrogen-reducing methods include using a secondary carbon source for the anoxic chamber and/or the use of multiple chambers[6].

One alternative is to replace the recirculating upflow filter with a second chamber that receives all of the nitrified wastewater. This second chamber, which consists of a deep sand filter, provides an anoxic environment to denitrify all of the wastewater in the system. Because the nitrified effluent will also have a low BOD, a carbon source must be provided to feed the bacteria. Methanol, a simple alcohol, can be used. Typically, the methanol to nitrate dosing will be 3:1. The effluent is then discharged from the system[7,8].

By increasing the sophistication of the design, it is possible to raise the removal efficiency to between 85 and 95 percent. Because the system is designed for bacterial growth in the sand, the filter will occasionally require backwashing to remove the excess bacterial growth. Figure 11.2 shows a second chamber system.

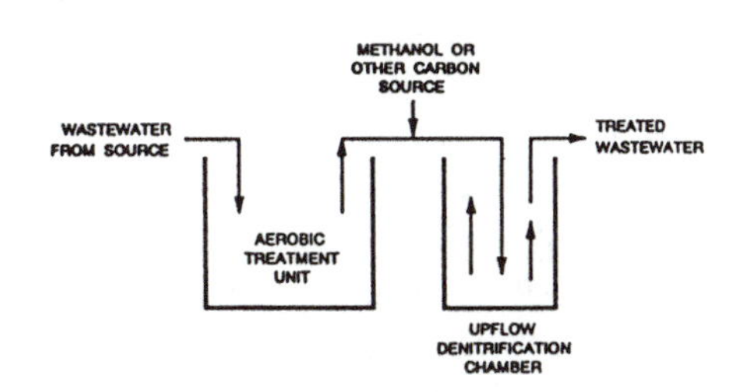

Figure 11.2
Second Chamber Design

Another variation that is used on a municipal scale is the four-stage Bardenpho system, as shown in Figure 11.3. Like the upflow anaerobic filter, the Bardenpho system uses carbon in the wastewater to feed the facultative bacteria and recycling to promote denitrification. Unlike the upflow filter, the Bardenpho system uses stirred tanks, not sand for this treatment. The bacterial cells are recycled, and the wastewater passes through alternating cell tanks that are kept in either aerobic or anoxic states[9].

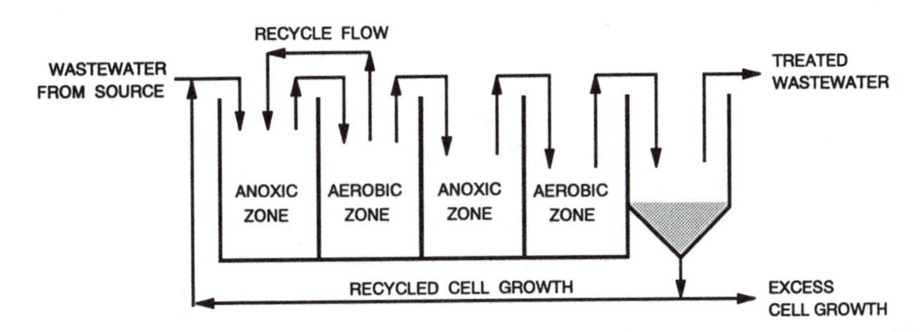

Figure 11.3 - Bardenpho System

Another system used on a municipal scale is the oxidation ditch, as shown in Figure 11.4. This system, which relies on a "racetrack" (looped) configuration, operates by passing the wastewater through an endless loop of aerobic and anoxic states. This system, while effective, does not typically have the same removal efficiency as the Bardenpho system.

Plant Uptake. As an alternative to treatment, the soil absorption can be designed, within limits, to use turf to remove the nitrogen. Essentially, this technique relies on a disposal system that discharges the effluent to the root zone. The discharge and nitrogen loading, to be effective, must be within the water and nutrient requirements of the plant. Plants that receive too much water will die, excess nitrogen wasted. Nitrogen uptake is limited to the growing season, so no nitrogen will be removed during the dormant season[11].

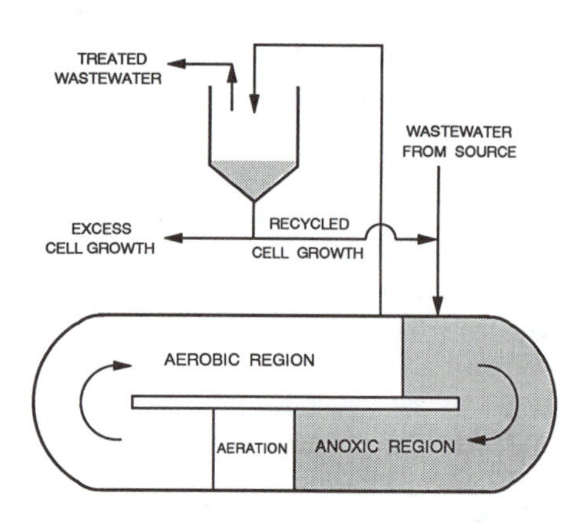

Figure 11.4 - Oxidation Ditch

The method of irrigating the soil must be carefully considered when using plant uptake for removal of nitrogen. Overland flow or spray irrigation is often subject to many local regulations, due to the possibility of disease transmission by aerosols or by ponded effluent. The other method, subsurface irrigation, is discussed in detail in Chapter 7.

Physical/Chemical Nitrogen Removal. Three methods of physical/chemical nitrogen removal are generally available and may be feasible under proper circumstances: breakpoint chlorination, ion exchange and air stripping. All three methods convert all of the nitrogen compounds to ammonia, then treat the ammonia. The BOD and TSS in the wastewater must first be treated before nitrogen conversion can take place. Physical/Chemical systems to remove nitrogen in onsite systems are impractical unless they can be demonstrated as sufficiently reliable and affordable for individual households[12].

Breakpoint chlorination is the addition of sufficient chlorine to the wastewater to oxidize all of the ammonia to nitrogen gas according to the following reaction:

$$NH_4^{+1} + 1.5\ HOCl \longrightarrow 0.5\ N_2 + 1.5\ H_2O + 1.5\ Cl^{-1}$$
$$2NH_4^{+1}OH^{-1} + 3\ NaOCl \longrightarrow N_2 + 5H_2O + 3\ Na^{+1}Cl^{-1}$$

This process provides for 99 percent nitrogen removal if up to 10 mg of chlorine are added for each mg of ammonia in the wastewater. Some of the untreated nitrogen will be in the form of nitrate and nitrogen trichloride (NCl_3).

This approach has obvious disadvantages: the storage of chlorine onsite and production of chloride, itself subject to groundwater protection standards, as a byproduct. Also, this type of process requires continuous monitoring and maintenance.

Selective ion exchange is possible through the use of the zeolite *clinophlolite*, which is contained as an ion exchanger. In this system, treated (but not nitrified) wastewater is passes through it. This type of system can achieve up to 97 percent removal. Sodium chloride is used to regenerate the system, just as in chlorination, ion exchange produces chloride as a byproduct. Alternatively, calcium or sodium hydroxide (CaOH or NaOH) can be used to regenerate the ion exchange bed. If a hydroxide is used, the volume of precipitate produced could pose operational problems and handling of these highly-corrosive chemicals poses safety risks[13].

Air stripping is the third physical/chemical method of nitrogen removal. In this process, the treated (but not nitrified) wastewater is raised to a pH between 10.5 and 11.5. At this pH, the solubility of ammonia in water will be reduced over 90 percent. The ammonia is actually removed by pumping the wastewater through a tower that has a large surface area. The surface area can be created by using many stacked trays of wood or synthetic media. The ammonia will diffuse to the atmosphere in accordance with Henry's Law. Henry's Law, after William Henry (1774-1836) states that the solubility of a gas in a liquid is directly proportional to the partial pressure of the gas above the solution in a closed container. So, as the pressure on the gas is reduced by exposing it to the atmosphere, the solubility of the nitrogen gas in the effluent is reduced, so it escapes to the atmosphere.

Operational problems with the unit include the need to add acid periodically to lower the pH, deposition of minerals on the media and tray surfaces and freezing of the units during cold weather.

Phosphorous Removal

Phosphorous removal is an emerging concern with regard to wastewater treatment because phosphorous is often the *limiting nutrient* for excessive algae and plant growth in surface waters. The role of this nutrient in the ecosystem is discussed on page 33.

Phosphorous is less of an environmental issue because most soils serve as a "sink" for phosphorous. This sink can be almost infinite because phosphorous concentrations in wastewater tend to be low, about 8 mg/L, and the adsorption capacity of the soil is high. The combining capacity is high because soil minerals preferentially combine with the phosphate ion in the form of orthophosphate (PO_4^{-3}), especially when calcium-based minerals are present. Phosphorous removal may not occur in sandier soils and where surface water discharge is allowed. For this reason, phosphorous removal may be required if a sensitive waterbody, such as a lake or stream, is nearby. Orthophosphate levels in the order of 25 ppb (parts per billion) can cause eutrophication of waterbodies, so preventing phosphates from entering waterbodies is essential to good environmental practice.

Some phosphorous treatment occurs during the wastewater treatment process. In a typical municipal wastewater treatment process, up to 30 percent of the phosphorous can be removed by bacterial assimilation. For example, if the influent concentration is 8 mg/L, the effluent phosphorous concentration would be over 5 mg/L. But, to minimize eutrophication, the effluent phosphorous concentration should be below 1 mg/L (1,000 ppb).

Phosphorous removal processes falls into three categories: biological removal, chemical precipitation, or plant uptake.

Biological Removal. Biological removal of phosphorous occurs through a phenomena called *luxury uptake*. In the process, the phosphorous-seeking bacteria are stressed through exposure to anaerobic conditions. The stressed bacteria are then exposed to the phosphorous-laden wastewater and aerobic conditions. In response to the stress and exposure, the bacteria will ingest more phosphorous than is necessary to meet their nutrient requirements. If the bacteria are then removed from the system, the phosphorous removal can exceed 80 percent. The challenge to the designer is to control the process so to remove bacteria from the system before they release the excess phosphorous ingested. If the bacteria are not removed and re-exposed to anaerobic conditions, they will release all the phosphorous they ingested. For an individual onsite wastewater treatment system, biological removal would be difficult to control[14].

There are alternative designs based on this phenomenon. For example, the Bardenpho process is a method of removing both phosphorous and nitrogen because the wastewater is exposed, to anaerobic, anoxic, and aerobic stages. Another process, known as a Sequencing Batch Reactor (SBR) is being examined for use in onsite systems. The SBR process, described in Chapter 6, uses a single tank in which various conditions are created through the use of pumps, timers, and control devices. The general sequence for phosphorous removal is: The tank is filled with a mixture of bacteria and wastewater. The contents are then mixed without the addition of oxygen so that anaerobic fermentation and release of phosphorous will occurs. The tank is then aerated, resulting in luxury uptake of phosphorous. After a certain time, aeration stops and the bacteria settle to the bottom of the tank and the clarified liquid is pumped out. Figure 11.5 shows the elements of various types of phosphorus removal systems[15,16].

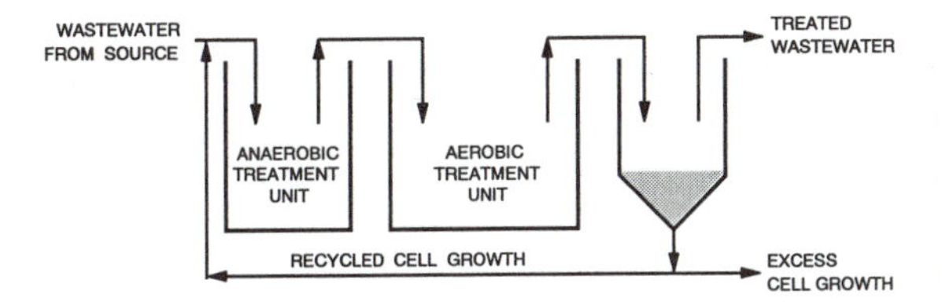

Figure 11.5
Phosphorus Removal Systems

While biological treatment is difficult for an individual owner, phosphorous uptake by plants is an alternative treatment method. As with nitrogen removal, plant uptake of phosphorous depends on the water and nutrient requirements of the plants and the climate.

Physical/Chemical Phosphorous Removal. The most common phosphorous removal method is the use of precipitating agents, principally iron, aluminum, and calcium salts. A fourth substance, "pickle liquor," which is a byproduct of steel production, is used in localities where steel production occurs. The specific chemicals used are lime ($Ca(OH)_2$), alum ($Al_2(SO_4)_3 \cdot 18\ H_2O$), iron(III) chloride (ferric chloride, $FeCl_3$), or iron(II) chloride (ferrous chloride, $FeCl_2$). Pickle liquor is an iron salt in a solution of sulfuric or hydrochloric acid. Using chemical precipitation, phosphorous removal exceeding 90 percent is possible.

A typical treatment scheme for the greatest phosphorous removal would include a phosphorous-removal unit as a third (or tertiary) treatment unit, after a septic tank and an aerobic treatment device. The tank is dosed with the metal salt and mixed with the treated wastewater. The resulting insoluble phosphate salt precipitates to the bottom of the tank as sludge. This sludge can be drawn off at the bottom while the clarified liquid is drawn off at the top. A chemical process to remove phosphorus is shown as Figure 11.6

While simple in concept, in practice, chemical precipitation is quite sensitive in practice. First, since precipitation is pH-dependent, the pH of the wastewater must be at an optimum pH, which is different for every metal salt. The pH will have to be adjusted to reach this point if not at there already. Second, the reaction is not limited to phosphorous, but will react with other anions (negatively-charged ions) in the wastewater and form other precipitants. As a result, the dose of metal salt and volume of total precipitants formed must be calculated and factored into the design if phosphorous removal is to be effective. Third, a management system for the unit is a necessary to ensure that both dosing and precipitant removal are maintained.

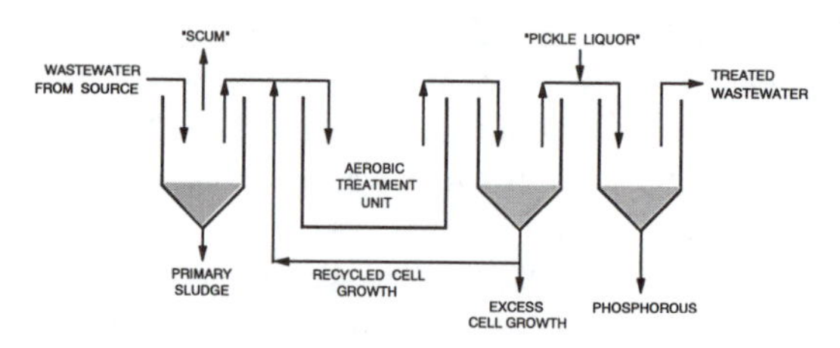

Figure 11.6
Chemical Process Design

GROUNDWATER MODELLING

If an onsite system is working "properly," that is, not creating a nuisance, contaminating a well, or discharging into a surface water, scant attention is paid to the where the water goes and what might be in it. Many homeowners assume that the effluent percolates through the soil and is magically rejuvenated back into pure drinking water. According to this view, effluent from the onsite system can be seen as water discharging into a pool where the water is mixed until it is no different from the water in the pool.

This view can even find its way into planning tools. The National Association of Home Builders has developed a planning and educational tool based on this assumption. This tool, which is intended to calculate nitrogen loading from a subdivision, is based on the premise that the nitrogen beneath a subdivision is the quotient of the mass of nitrogen from onsite systems (and other sources) divided by the mass of water[17].

The idea that mixing beneath a site occurs might appear reasonable, but such is not the case. Rather than mixing, effluent from onsite systems is marked by a characteristic lack of mixing.

Plumes

As discussed in Chapter 5, *Hydraulics*, flow in soil is essentially laminar and can be estimated using the Darcy Equation. Laminar flow, as discussed earlier, is characterized by a lack of turbulence and by flow streams of equal velocity. Effluent is also characterized by concentrations of dissolved, sodium, chloride, and nitrogen, that will most likely be different from the natural water. Because the effluent has a different characteristic, it will have a different density. As a result of the laminar flow and different density, effluent from onsite systems will not readily mix with groundwater.

Cherry, et. al. investigated effluent movement at two sites in Ontario. One site had an onsite system in use for 12 years; the other had an onsite system in use for 1.5 years. Their results, which are consistent with other research, verified that effluent movement occurs in plumes that show little attenuation for a significant distance. Effluent movement was estimated to be 40 m/yr (131 ft/yr). At this velocity and in the sandy soils characteristic of the sites, the authors concluded that the plume having an initial concentration of 33 mg/L-N would have to be over 170 m (557 ft) to meet the drinking water standard of 10 mg/L-N and be 2 km (6,561 ft) long to achieve a concentration of 2.5 mg/L-N, which is a concentration used to initiate remedial action. Figure 11.7 shows a plume from an onsite system[18].

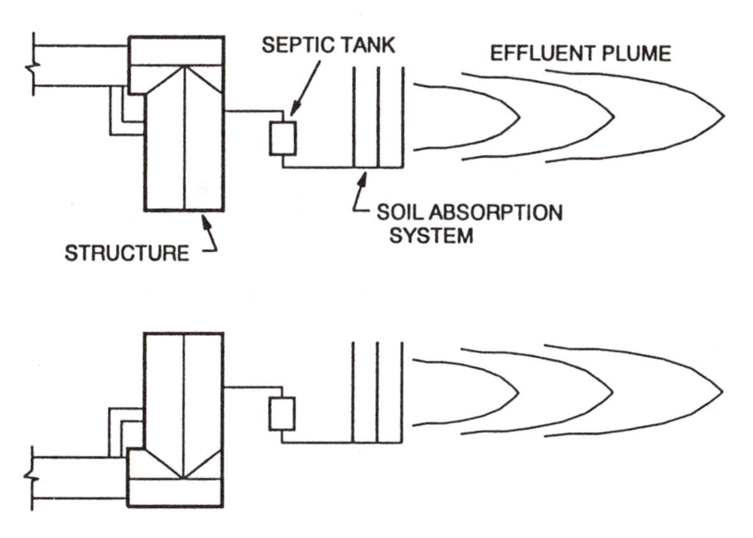

Figure 11.7 - Effluent Plumes from an Onsite System

The authors' conclusion is that current separation distances between onsite systems and wells are probably inadequate depending on the relative locations of the wells and onsite systems. Going one step further, the EPA designated areas having more than 16 onsite systems per km^2 (40 onsite systems per mi^2) as regions of potential groundwater contamination[19].

SUMMARY

Onsite systems are viewed as a source of groundwater contamination for waterborne diseases and nutrients. Treatment strategies to minimize the risk of waterborne diseases were discussed in Chapters 6 and 7. The designer must also address environmental issues related to nutrient removal, the density of onsite systems, and the movement of groundwater from the soil absorption system. As an alternative, the designer may consider treatment units, principally for nitrogen and phosphorous removal. Regardless, the designer will be able to choose among alternative systems that provide the degree of environmental protection being demanded of onsite systems.

References

1. Sedlak, R., Ed. *Phosphorous and Nitrogen Removal from Municipal Wastewater: Principles and Practice*. New York: The Soap and Detergent Association, 1991, pp 3-7.

2. Fan, A., Willhite, C., and Book, S. "Evaluation of the Nitrate Drinking Water Standard with Reference to Infant Methemoglobinemia and Potential Reproductive Toxicity." *Regulatory Toxicology and Pharmacology*, Vol. 7, pp 135-148.

3. Henry, J.G. and Gary W. Heinke. *Environmental Science and Engineering*, Prentice-Hall, Englewood Cliffs, NJ, 1989, p. 290.

4. Siegrist, R. and Jenssen, P. "Nitrogen Removal During Wastewater Infiltration as Affected by Design and Environmental Factors." Madison: Ayres, Associates, 1989.

5. Mitchell, D. "Laboratory and Prototype Onsite Denitrification by an Anaerobic-Aerobic Fixed Film System." Fayetteville: University of Arkansas, Department of Civil Engineering, 1989.

6. Sedlak, p 51.

7. Sedlak, pp 46-48.

8. U.S. Environmental Protection Agency. *Onsite Wastewater Treatment and Disposal Systems*. EPA 625/1-80-012. Washington D.C.: Office of Water Program Operations, 1980, pp 188-190.

9. Sedlak, pp 48-49

10. Laak, R. *Wastewater Engineering for Unsewered Areas*. Lancaster: Technomic Publishing Co., Inc., 1986, p 42.

11. U.S. Environmental Protection Agency. *Process Design Manual for Land Treatment of Municipal Wastewater*, EPA 625/9-76-0003.

12. Tchobanoglous, G. "Physical and Chemical Processes for Nitrogen Removal: Theory and Application. Proc. 12th Sanitary Engineering Conf, University of Ill, Urbana, 1970.

13. Sedlak, p 40.

14. Burks, B. *Phosphorous Removal Using the Bardenpho Process*. Unpublished Paper.

15. U.S. Environmental Protection Agency. *Phosphorous Removal*, EPA/625/1-87/001. Cincinnati: Center for Environmental Research Information, 1987.

16. Sedlak, pp 142-178.

17. "A Procedure to Determine Optimum Density for Homes Using Individual Wastewater Treatment Systems Based on Nitrogen in Groundwater Recharge." Upper Marlboro: NAHB National Research Center, 1989.

18. Cherry, J., Robertson, W, and Sudicky, E. "Ground-water Contamination from Two Small Septic System on Sand Aquifers." *Groundwater*, Vol. 29, No. 1, pp 82-92.

19. Yates, M. "Septic Tank Density and Ground-water Contamination." *Groundwater*, Vol. 23, No. 5, pp 586-591.

Paul Ashburn of Morgantown, West Virginia, has found an innovative method to expand the performance capability of his custom-designed sand filters. During the growing season, he plants flowering vegetation in and around the sand filters to enhance evapotranspiration and nutrient uptake.

CHAPTER 12

CONSIDERATIONS FOR ONSITE SYSTEM USE

Topics of This Chapter

- *Issues Surrounding OWTS Use*
- *Daily Implementation of onsite system Programs*

INTRODUCTION

The selection and design of an onsite system might seem to be a straightforward technical matter based on principles of soil science, chemistry, engineering and biology. A reader of this book might decide that he/she was ready to design and install an onsite system. This may not be the case because the regulation of onsite systems in many regions is tightly bound to plumbing, land use, or environmental regulations that seek to use the prescriptions of onsite system regulations to achieve other goals. The regulation of onsite systems, rather than being a technical issue, is often a political matter in which technical questions are not addressed until political issues are settled.

This chapter will provide information about three of the issues that may frustrate or enhance the selection and installation of onsite systems. These issues are separate, but interrelated if only because they are used to bolster or supplant each other, depending on the local or regional politics.

LAND USE MANAGEMENT

Land use management is loosely defined as the orderly division and development of parcels in accordance with a land use plan. This plan can be a formal document that encompasses the management of all land use. For example, many cities and metropolitan regions have specific planning departments or planning commissions whose purpose is to manage the growth and land use of the jurisdiction. Land use planning may not be the result of a comprehensive plan; it may result as an outgrowth of a specific planning effort. Many municipalities experienced this when they upgraded their municipal wastewater treatment facilities. As a part of the renovation process, the governing agency was required to identify a service area for the wastewater treatment facility. This service area, known as the "201 planning area" (which is derived from the Clean Water Act), sets the boundaries for the installation of municipal sewers.

In most areas there is no formal plan; "planning" is simply an extension of the predominant political beliefs of the constabulary. For example, in one region of the country there is a predominant belief that "a man's home is his castle." If this is the local attitude, meaning that property owners are sovereign over their properties, then land use management is really the absence of land use planning. Each owner is "free" to use land as deemed appropriate, the only limitations imposed by economic consideration or dangers posed to the neighbors. Until recently, Houston, Texas, had no zoning regulations, which led to some interesting land-use mixtures in the area, like businesses among residences.

The other extreme of land use planning was in the former totalitarian government in Romania, where "land use management" meant the destruction of entire villages and forced move of entire populations to cities where citizens' activities could be more closely monitored. The American version is not as brutal, but can feel very restrictive to some. This example can be seen in "planned communities" where all aspects of lot development—including architectural features, style, size, height, and a myriad of other details—are closely regulated for conformance with the restrictions outlined in covenants attached to the deed of each lot.

Most of the United States falls is a funciton of the end of sewer pipes, condition of the local wastewater treatment plant, local zoning ordinances, administrative rules governing construction, and deed restrictions for lots within subdivisions.

Development outside of areas served by sewers is a particularly nettlesome issue for land use managers. Sewered development is limited to the lengths of the sewer pipes, a feature that has limitations. Moreover, planners often hope that the availability of sewers will result in "infilling"— the use, or re-use, of vacant lots in sewered areas.

People want to live in unsewered areas for various reasons. Some want to live on larger parcels where the cost of housing is lower. Others want to escape the perceived evils of urban life. However, as rural property is developed, many municipal services follow the population. Soon office complexes spring up to eliminate long commutes, and roads and utilities are built in response to the population's needs. The end result of this cycle is the scattered development known as "urban sprawl."

Opponents of urban sprawl contend that development in sewered areas, which are controlled, protects society better. Forcing people to remain in sewered areas has several distinct advantages. It protects cities from an erosion of the tax base. It also would lower the costs to operate wastewater collection and treatment facilities because existing systems are maintained, not abandoned in favor of newer systems to service scattered developments. Utility costs are lower because services would not have extended to outlying districts. Further, environmental impacts are minimal. There is no pollution from increased vehicle use—automobile emissions and highway construction. There is no need to damage environmentally sensitive area to expand municipal services such as water mains and electric power.

Onsite systems are often blamed for urban sprawl because many believe that development could not occur if onsite systems were not available. The corollary argument is that onsite systems are sometimes considered a "temporary" wastewater management measure, destined to fail. Therefore, municipal sewers must be constructed eventually, thus repeating the cycle of sprawl as even more development occurs along the fringe of the newly-expanded sewer service area.

Use of Regulations

Given the differing approaches to land use management, it is not surprising that groups in the political factions try to exert control through alternative means especially when they haven't sufficient support to exert political control. Administrative rules are one such vehicle, particularly for those who oppose urban sprawl. They provide several advantages that favor those who want to exercise greater control:

- Difficult to Change
- Conservative in terms of being restrictive
- Often supported by appeal to "Mom and Apple Pie" emotions
- Protect favored interest groups

Onsite system regulations are particularly susceptible to all of these political tactics. Most onsite system designs rely on soil infiltration to operate. Therefore, the extent of onsite system use, therefore unsewered development, can be limited by the soils in an area. For this reason, onsite system regulations have been a favorite target for use as a *de facto* land use management tool in the absence of local land use planning.

Difficult to Change. Onsite system regulations are typically embodied in a ordinance or administrative rule. *Ordinances* are laws passed by a county or one if its political subdivisions. They are introduced by members of the governmental unit, such as a county board member or city councilman. They will be debated by the board or council, which will receive public testimony before voting on the issue. If adopted, the ordinances become the laws of the community.

Administrative rules are regulations adopted by a state regulatory agency in response to state statutes. The state statute will address a specific issue, such as public health or environmental protection, and direct a state agency to adopt rules that carry out the intent of the statute. Administrative rules are then developed by that agency, usually based on input from an advisory committee and after a round of public hearings. The rules are then submitted to the state legislature, which can object to the rules within a specified time period, say 30 days. If there are no objections, the agency is free to promulgate the rules though the official journal of the state and then enforce them[1].

The process for amending rules is the same as for promulgating new rules. The amendment must be developed, taken to public hearing, submitted to the legislature, and promulgated by the agency before it can take effect. Figure 12.1 shows a typical meeting of administrators.

Figure 12.1 - Administrative meeting

Some states require rules (or amendments to existing rules as shown in Figure 12.2) potentially having an environmental impact to be subject to an *environmental assessment*. If the assessment reveals that the rules have a significant environmental impact, the agency may be required to develop and distribute an *environmental impact statement*, which is a formal document detailing the type of environmental impacts the proposed rules may have and the extent of the impacts.

The administrative rule process is involved and lengthy. Assuming there are no problems, it takes six months from the time a rule is proposed until it can be submitted to the legislature. If the rule is controversial or requires an environmental assessment, the time period may be lengthened to a year or more for the time to complete additional public hearings on the rule and environmental assessment. Then, the legislature may object to the rules—either rejecting them, conducting additional hearings, or suspending the rules after they are promulgated.

Conservative in terms of being restrictive. Because the administrative rule process can be so tortuous, administrative rules are difficult to change. First, the agency must be inclined to develop new rules or amend existing rules. Agencies will not usually change rules unless they see a benefit to the change. Agencies, like other monopolies, are predisposed to propose rules that make their job easier and more secure, regardless of the effect on the citizens.

Rules that are management challenges or "risky" are avoided. "Risky" in this sense can include rules that include new technology, changes to time-proven methods, or amendments that irritate favored constituencies of the agency.

Wisconsin Administrative Rules

Uniform Plumbing Code

Private Sewage Systems

Chapter COMM - 83

Figure 12.2
Administrative Rules

Appeal to "Mom and Apple Pie" Emotions. State government exists to protect the health, safety, and welfare of its citizens. Of these three missions, *welfare* can be the most difficult to define and administer. Welfare issues can include everything from "consumer protection" to "clean environment," the exact definition of each is developed by the agency staff. As a result, it is easy for emotional appeals to interfere with the review of scientific data. No politician or regulator wants to be accused of being insensitive to environmental needs. How can a regulator support a rule that may endanger the citizens, no matter how minuscule the risk? What agency wants to be accused of allowing groundwater to be polluted and then poison innocent babies? Such appeals cause the agencies into highly defensive opinions, particularly in the face of a vocal constituency, no matter how small.

Protected by Favored Interest Groups. Rules that favor protected trade or environmental groups are also jealousy guarded by the these groups. Plumbers accustomed to skilled, labor-intensive activities have often opposed labor-saving innovations. Likewise, they may oppose technologies that are unfamiliar or which favor competing trades. Environmental groups will oppose rules that have adverse environmental impacts, as the group has defined the phrase "adverse environmental impact." Trade groups, by contrast, will oppose regulations that require them to upgrade the performance of their products or that may provide a competitive advantage to products they don't manufacture.

Onsite system rules suffer from all of these. Many rules are tradition-bound, rooted in decades-old practices. The rules protect an entrenched industry that is comfortable with time-honored methods and materials of construction. This also favors the regulators, since administration of the rules is easy because all involved with the process are familiar with the process. Because the rules have been in effect so long, they recognize technologies that seem to work; there is little chance the agency or industry will suffer the embarrassment of or retribution from failing technology.

Environmental and public health concerns can be the most difficult to address. Traditional onsite systems address acute public health issues related to exposure to raw wastewater. Increasingly, onsite systems must address chronic public health concerns and environmental issues. These new concerns and issues can immobilize an agency that has neither the resources nor will to address them.

The Metropolitan Milwaukee Experience

The Milwaukee metropolitan area encompasses seven counties that covers 2,689 square miles, includes 154 local governmental units, and is home to about 40 percent of Wisconsin's population. Before World War II, the region had one of the densest populations in the United States, at 11,346 people/mi^2 (29,386 people/km^2). By the 1960's population density had plummeted to about 3,600 people/mi^2 (9,324 people/km^2). In response to this trend and the associated problems, area leaders lobbied for the authority to create a planning commission, which was formed in 1960 and named the Southeastern Wisconsin Regional Planning Commission (SEWRPC). SEWRPC was given the mission of developing a coordinated land use management plan for the region, and in 1966 the staff produced a three-volume plan[2].

SEWRPC developed a plan based on growth contiguous to already developed areas and using wastewater conveyance and centralized wastewater treatment. The theory behind this concept was that the region should maintain a dense population, and only centralized wastewater collection and treatment could provide for such density[3].

SEWRPC planned for development to occur where sanitary sewers were available and that the expansion of sanitary sewers would be in accordance with the land use plan. As a corollary of this plan, SEWRPC identified areas it assumed would be unsuitable for development based on onsite system regulations at the time. SEWRPC held the belief that the siting requirements in the code would be unchanging, so a plan based on these requirements would be unchanging. Land use plans were developed using USDA-SCS soil maps.

Figure 12.3 - 1963 aerial photograph of a rural Wisconsin area.

At the same time however, the University of Wisconsin was engaged in research to identify an alternative onsite system that could address the very soil and siting limitations experienced in the Milwaukee area. As a result of their efforts, researchers developed the Wisconsin mound system, which is essentially a single pass sand filter constructed over and integrated with native soil.

The mound system had the effect of invalidating the SEWRPC plan because the mound system could be used on sites undevelopable using traditional septic systems. Because there was no consensus on the master plan, development began in sites identified as undevelopable under the original plan[4].

Nor could SEWRPC force its plan upon local governments. They were free to adopt the regional plan as they pleased, and many chose to ignore it in the race to ease property tax burdens by encouraging development. Indeed, some local government officials took views opposite to SEWRPC, seeing their responsibility to promote development within their respective jurisdictions. As a consequence, some governments adopted the regional plan even as others ignored it.

Figure 12.4 - The same area as that shown in Figure 12.3, approximately thirty years later. Note that much of the former farmland has been subdivided for residences.

SEWRPC may have criticized local governments for failing to follow the "accepted" plan, it did criticize the state agency responsible for the plumbing code, claiming the agency had abrogated its responsibility to recognize the land use impacts of its decisions[5]. The staff complained that state agencies should develop uniform state polices for encouraging compact, contiguous urban growth. Taking this process one step further, the staff recommended that development be allowed only in accordance with approved state-level plans for wastewater management. Development using onsite wastewater management would be effectively prohibited if the state-level plan did not recognize it. And, the staff recommended that a state agency long antagonistic to onsite wastewater management be the one to approve the wastewater management plan.

SEWRPC was joined in its criticism and recommendations by those who argued that land use impacts are part of the general environmental impacts of onsite systems. All environmental impacts of these systems, particularly in the case of newer technology, must be assessed before the technology is allowed. Further, technology must be restricted if the technology has adverse secondary impacts, even if the technology is itself beneficial. In fact, onsite wastewater management should be banned because such technology is considered a temporary, failure-prone technology[6,7,8].

SEWRPC's argument is not held by all. There is another view many subscribe to: onsite system technology should be developed independent of land use considerations. Land use management should be discussed solely within the confines of land use management plans, not onsite system codes. In this view, land use is a local matter to be decided and regulated through zoning ordinances[9].

Zoning is a process of allocating land development and land use in accordance with local zoning ordinances, which must be enacted by local governmental units. The zoning process either results from the codification of existing practices, or it can result from the planned development of a specific region. The SEWRPC plan, for example, recommended that agricultural areas be zoned exclusively for agricultural use. The reason for the recommendation, which was adopted by some governmental units, was two-fold. First, zoning would prevent conversion of agricultural land, all of it quite fertile, into low density housing units. Second, zoning would, at least in theory, allow family farms, a hallmark of Wisconsin, to remain in perpetuity.

Zoning also allows a government to designate certain areas for specific activities, whether it is residential, commercial, or industrial use. Each use, which has its own specific transportation and utility requirements, can be serviced individually and efficiently within designated zones. In addition, zones can be selected so that they are not in conflict with each other. Zoning, proponents argue, should be used as the sole method for regulating land use, and this zoning should be developed in accordance with land use plans[10]. The question, remains: Whose land use plan should be followed?

Generalized Experience

The Metropolitan Milwaukee experience has been repeated elsewhere. Wastewater management is often the most sensitive of the issue for any development. The availability of other utilities such as roads, water, power, and telephones, is seldom questioned. Onsite system technology is under intense scrutiny because it affects the sensitivity of site development. For this reason, onsite system technology has been linked to land use. In the absence of land use planning, the control of onsite system technology can be the most important vehicle for controlling development.

Onsite system technology is subject to extensive examination in every jurisdiction where it might be used, and because technologies must be approved before they are used, approval must be granted in each locality that has the authority to promulgate onsite system codes.

Newer technologies must compete against traditional designs that are considered relatively "failure-proof," meaning the designs require little maintenance and are based on traditions that may include a multiplicity of safety factors. The compounding factors of safety result in an almost absolute assurance of proper functioning regardless of use or abuse. Newer technologies often require more maintenance and have fewer of the safety factors built into the system design. They are, therefore, more prone to failure if abused or neglected[11].

Implementation of Onsite System Regulations

In the typical community, onsite regulations are enforced at the county level, typically by the county sanitarian, public health department, or planning department. The choice of agency is somewhat arbitrary, depending on the how local officials organize county government. Regardless of agency, the function of the staff will be to ensure that onsite systems are designed, installed, and operated in accordance with applicable statutes, ordinances, and rules.

Some jurisdictions provide a greater degree of regulation than others. For example, some jurisdictions provide staff who conduct soil and site evaluations, then develop designs from them. Other jurisdictions inspect the services provided by private soil evaluators and designers. Some jurisdictions allow a homeowner to install an onsite system of their own design. Other jurisdictions have elaborate systems of design, plan review, and installation, all by properly licensed or certified personnel.

Site Evaluation. The process starts with an owner who wants to develop a particular site. The owner will either contact the local public health official or licensed professional, who will evaluate the suitability of the site for an onsite system. This evaluation will either be conducted or inspected by the public health officials. The evaluation and/or inspection may include soil borings, a review of soil maps, or logs of other inspections in the same area. Regardless, the site may not be considered acceptable until after it has been judged suitable by local officials.

System Design. After the site is evaluated, a onsite system must be designed to meet any soil and site restrictions there may be. As noted, this design may be conducted by local officials, or it may be executed by an installer, engineer, plumber, or other person qualified by license, certificate, or experience to do so. If the design is executed by a service provider, plan review and approval may be required before the system can be installed. The approval usually involves the development of plans and specifications, and the review will focus on conformance with code requirements, particularly setbacks, tank capacity, and soil absorption area.

Permits. Once the plan is approved, the owner may apply for various permits, including one to install the onsite system, perhaps another to begin construction of the structure, another to install a well. There may be other permits as well. In fact, the ability to obtain other permits may hinge on the possession of a permit to install a system. That is because different permits may be issued by different local or state agencies, each of which may need to see the permit or permits issued by other agencies.

Some jurisdictions absolutely require that a permit to install an onsite system be obtained before any other permits. This permit, which might be called a *sanitary permit*, serves two purposes. First, it ensures that the sanitarian has time to have the owner protect the intended onsite system site from damage during the construction of the structure. Onsite system areas, particularly for above grade systems, are vulnerable to damage by construction equipment and must be protected during construction. The sanitary permit serves as a notice that the premises will be served by an onsite system. Jurisdictions usually have appeal processes, and neighbors may object to the issuance of the permit. The notice gives them an opportunity to challenge the issuance of the permit.

Installation. Once the necessary permits are obtained, the installer is ready to construct the onsite system. In many jurisdictions, this may be the most scrutinized step. The construction may be inspected up to seven times depending on the complexity of the system. For example, mound systems are often inspected after the vegetation is removed, after the soil is plowed, after the sand is installed, after the pressure distribution system and tanks are installed, and after the soil cap is placed over the mound. The mound may be inspected one final time after it is put into service to ensure that pumps and alarms are properly installed and functioning.

Occupancy. The owners may be required to obtain an occupancy permit before they are allowed to occupy the structure. The occupancy permit may include final checks that the structure, including the onsite system, has been inspected as required and that all fees have been paid. Only then may the owners inhabit the structure and use the onsite system.

Ongoing Maintenance. Some jurisdictions require that owners of onsite systems, particularly those using mechanical systems, purchase a service contract with someone qualified to maintain the system. For newer fabricated aerobic treatment units, the manufacturer specifies a service provider, who is probably the installer. Generally too, the installer will be the service provider.

Service for most onsite systems is a quarterly or semiannual activity. The condition of the pumps, filters, and other units will be inspected and cleaned or replaced as necessary. Sometimes, effluent quality samples will be taken for inspection. In most cases, the unit is assumed to be functioning properly if it is operated and maintained in accordance with the manufacturer's instructions.

Enforcement Activities. In addition to inspecting onsite systems when they are installed, public health officers must also enforce regulations to prevent and abate public health hazards that could result from an improperly operating onsite system. At a minimum, an improperly operating onsite system creates a nuisance, either with odors or water. Worse, an improperly operating system could discharge pollutants to nearby streams or groundwater. Wastewater on the ground can infect people who step in it or serve as a reservoir for vermin.

Enforcement activities can occur from a complaint, as part of routine inspections, or during a specified activity such as the transfer of the property or application for a building permit. The activity can be as simple as verifying that the onsite system conforms to applicable codes and is properly sized for intended use. This provision generally applies to older onsite systems whose condition is unknown or was installed before accurate records were kept at the sanitarian's office. Or, in the case of an expansion or change of use, the action is intended to ensure that the onsite system has sufficient capacity to treat the wastewater generated after the expansion or change in use has occurred. If the onsite system is insufficiently sized, the owner must enlarge the system before the expansion or change in use begins. And, the site of the enlarged system can be protected during the expansion or change in use.

In some cases, the onsite system will be failing, that is, discharging untreated wastewater to the ground surface, a receiving water body, or to groundwater. In these cases, the system must either be repaired or replaced. In the case of older onsite systems, replacement is the only viable alternative since the older system probably will not meet current code requirements. If this is the case, the sanitarian will probably issue an *enforcement order* requiring that the system be replaced within a specified time period. If the system is not replaced, the owner may be subject to a citation or other legal action.

SUMMARY

The installation and operation of onsite wastewater treatment systems is not as straightforward as this text might at first make it appear. At the level of public policy, debate rages over the role of onsite system technology in the development and perpetuation of urban sprawl. Often onsite system regulations, which are essentially technical specifications, become the focus of political forces that seek to use, or abuse, the specifications to achieve other political goals. These goals include the use or avoidance of certain onsite system technologies, promotion of preferred land management paradigms, or protection of affected constituencies. As a consequence, technical issues often become embroiled in political arguments that generally hinder both technical and political decisions.

On a day-to-day basis, the resulting regulations are enforced by local public health officers whose duty it is to ensure that the regulations are met. These officials will inspect sites for their suitability for an onsite system, review plans for proposed onsite systems, inspect onsite systems as they are installed, and examine the installed onsite system if there is an indication that it is not functioning properly. Because the installation of an onsite system is so important, a permit to install an onsite may be necessary before any other permit can be obtained.

References

1. Chapter 227, Wis. Stats.

2. *Recommended Regional Land Use and Transportation Plans, 1990*, Vol 1-3. Report No. 7. Waukesha: Southeastern Wisconsin Regional Planning Commission, 1966.

3. SEWRPC, Vol 3, pp 4-6.

4. *A Regional Land Use Plan and a Regional Transportation Plan for Southeastern Wisconsin-2000*, Vol 1, Inventory Findings. Waukesha: Southeastern Wisconsin Regional Planning Commission, 1975.

5. Cutler, R. "Past, Present, and Future of Regional Land Use Plan Implementation Study." Presentation to the SEWRPC Commissioners in West Bend, June 16, 1993.

6. SEWRPC, Vol 3, p 12.

7. Falk, K, et. al. Petition to Amend Chapter ILHR 85, Wisconsin Administrative Code, 1989.

8. Dawson, T. Testimony in *State of Wisconsin Public Intervenor vs. State of Wisconsin Department of Industry, Labor and Human Relations*, 1992.

9. Skornicka, C. Secretary of the Wisconsin Department of Industry, Labor and Human Relations Correspondence to the Mayor of Stevens Point, Wisconsin, 1993.

10. Cutler, p 11.

11. Venhuizen, D. Correspondence from Washington Island, WI, Demonstration Project for Recirculating Sand Filters.

RESOURCES

National Small Flows Clearinghouse 800-624-8301
West Virginia University
P.O. Box 6064
Morgantown, WV 26506-6064

Small Flows operates a help line to assist in finding information about reserach and articles about onsite systems. Their well-trained professional staff provides helpful answers on all aspects of onsite systems.

National Onsite Wastewater Recycling Association 800-966-2942
Fact sheets about septic systems and record-keeping forms, information about state and regional wastewater associations.

US GAO (202-512-6000) *fax* 301-258-4066
P.O. Box 6015
Gaithersburg, MD 20884-6015

The GAO published *Water Pollution - Information on the Use of Alternative Wastewater Treatment Systems, 9/94.* The publication number is **GAO/RCED-94-109.** The first copy of this report is free.

North Carolina Rural Communities Assistance Project, Inc.
P.O. Box 941
Pittsboro, NC 27312
919-542-7227 *fax* 919-542-2329
This organization published *A Guide to Wastewater Management for Small Communities in North Carolina.* Call or fax for price.

Contacting the authors:
Bennette D. Burks - 608-241-3447
Mary Margaret Minnis - 203-866-3006

APPENDIX A

This table was adapted from the *Vertical Separation Distance Requirements for Onsite Systems* prepared by the National Small Flows Clearinghouse in Morgantown, WV, May, 1991. This information should serve only as a rough guide and local regulations should be consulted before any assumptions are made about regulations in a particular county or state.

STATE	Vertical Separation Distance for Conventional Systems
Alabama	18" - bedrock and seasonal g.w. table
Alaska	
Arizona	5' to g.w. elevation, 4' to limiting factor
Akransas	4' - bedrock & seasonal g.w. table
California	varies with the county
Colorado	4' - bedrock & seasonal g.w. table
Connecticut	18" - maximum g.w. table, 4' to limiting factor
Delaware	3' to a limiting zone
Florida	42" to a limiting layer
Georgia	2' - to bedrock & seasonal g.w. table
Idaho	2' - 6' based on slope, acerage and soil design group
Illinois	4' - g.w. elevation, 4' creviced limestone
Indiana	2' from any horizon with a loading rate less than 0.25 gpd/ft^2
Iowa	3' - bedrock & seasonal g.w. table
Kansas	4' - bedrock & seasonal g.w. table
Kentucky	2' soil depth, 18" for special system design
Louisiana	2' - seasonal g.w. table, 4' - impervious zone
Maine	12 -24" based on soil conditions
Massachussetts	4'
Michigan	4' to limiting factor (check county rules)

STATE	Vertical Separation Distance for Conventional Systems
Minnesota	3' - bedrock & seasonal g.w. table
Missouri	2' to limiting factor
Montana	4' to limiting factor
Nebraska	4' - seasonal g.w. table
Nevada	2 - 4', based on acerage
New Hamp-shire	8' to bedrock, 4' seasonal g.w. table
New Jersey	2' - 9', based on soil suitability class (measured from existing ground surface)
New Mexico	4' - bedrock & seasonal g.w. table
New York	24" to limiting factor
Ohio	12" to limiting factor
Oregon	6" to limiting layer, 4' from permanent water table
Pennsylvania	4' - limiting zone
Puerto Rico	4' bedrock & seasonal g.w. table
Rhode Island	5' - bedrock, 3' seasonal g.w. table
South Dakota	4' - bedrock & seasonal g.w. table
South Carolina	6" seasonal g.w. table, 1' from bedrock
Tennessee	4' - bedrock & seasonal g.w. table
Utah	4' - bedrock, 2' -seasonal g.w. table
Vermont	4' - bedrock, 3' seasonal g.w. table
Washington	3'
West Virgina	3'
Wisconsin	3'
Wyoming	4'

Appendix B

Parameter for Testing	Jar	Volume (ml)	Preservation	Maximum holding time
Bacterial Tests				
Coliform, fecal and total	P,G	—	Cool, 4°C, 0.008% $Na_2S_2O_3$	6 hours
Fecal streptococci	P,G	—	Cool, 4°C, 0.008% $Na_2S_2O_3$	6 hours
Inorganic Tests				
Acidity	P,G	100	Cool, 4°C	14 days
Alkalinity	P,G	200	Cool, 4°C	14 days
Ammonia	P,G	400	Cool, 4°C, H_2SO_4 to pH <2	28 days
BOD	P,G	1,000	Cool, 4°C	48 hours
Bromide	P,G	100	None required	28 days
CBOD	P,G	1,000	Cool, 4°C	48 hours
COD	P,G	100	Cool, 4°C, H_2SO_4 to pH <2	28 days
Chloride	P,G	50	None required	28 days
Chloride, residual	P,G	500	None required	14 days
Color	P,G	500	Cool, 4°C	48 hours
Cyanide, total and amenable to chlorination	P,G	500	Cool, 4°C, NaOH to pH >12 0.6 grams of ascorbic acid	
Fluoride	P	300	None required	28 days
Hardness	P,G	100	HNO_3 to pH <2, H_2SO_4 to pH <2	6 months
pH	P,G	25	None require	Analyze immediately
Kjeldahl and organic nitrogen	P,G	500	Cool, 4°C, H_2SO_4 to pH <2	28 days
Metals				
Chromium (VI)	P,G	500	Cool, 4°C	24 hours
Mercury	P,G	500	HNO_3 to pH <2	28 days
Metals, except chromium and mercury	P,G	200	HNO_3 to pH <2	6 months

Parameter for Testing	Jar	Volume (ml)	Preservation	Maximum holding time
Nonmetals				
Nitrate	P,G	100	Cool, 4°C	48 hours
Nitrate nitrite	G	200	Cool, 4°C, H_2SO_4 to pH <2	28 days
Nitrite	P,G	100	Cool, 4°C	28 days
Oil and grease	P,G	1000	Cool, 4°C, H_2SO_4 to pH <2	48 hours
Organic carbon	G, BT	100	Cool, 4°C, HCl or HSO_4 to pH <2	Analyze immediately
Orthophosphate	G, BT	50	Filter immediately, Cool 4°C	8 hours
Oxygen, dissolved - probe	G only	300	None required	28 days
Oxygen, dissolved - Winkler	G	300	Fix on site and store in dark	48 hours
Phenols	P,G	500	Cool, 4°C, H_2SO_4 to pH <2	28 days
Phosphorus, elemental	P,G	50	Cool, 4°C	7 days
Phosphorus, total	P,G	50	Cool, 4°C, H_2SO_4 to pH <2	48 hours
Residue, total	P,G	100	Cool, 4°C	48 hours
Residue, filtrable	P,G	100	Cool, 4°C	7 days
Residue, settleable	P,G	1000	Cool, 4°C	48 hours
Residue, volatile	P	100	Cool, 4°C	7 days
Silica	P,G	50	Cool, 4°C	28 days
Specific conductance	P,G	500	Cool, 4°C	28 days
Sulfate	P,G	500	Cool, 4°C	28 days
Sulfide	P,G	500	Cool, 4°C. add zinc acetate plus NaOH to pH>9	7 days
Sulfite	P,G	50	None required	Analyze immediately
Surfactants	P,G	—	Cool, 4°C	48 hours
Temperature	P,G	1000	None required	Analyze immediately
Turbidity	P,G	100	Cool 4°C	48 hours

Parameter for Testing	Jar	Volume (ml)	Preservation	Maximum holding time
Organic Test				
Purgeable halocarbons	G,TLS	—	Cool, 4°C, 0.008% $Na_2S_2O_3$	14 days
Purgeable aromatic hydrocarbons	G,TLS	—	Cool, 4°C, 0.008% $Na_2S_2O_3$, HCl to pH = 2	14 days
Acrolein and acrylonitrile	G,TLC	—	Cool, 4°C, 0.008% $Na_2S_2O_3$, adjust pH to 4 to 5	14 days
Phenols	G,TLC	—	Cool, 4°C, 0.008% $Na_2S_2O_3$	7 days until extraction
Benzidines	G,TLC	—	Cool, 4°C, 0.008% $Na_2S_2O_3$	40 days after extraction
Phthalate esters	G,TLC	—	Cool, 4°C	7 days until extraction
Nitrosamines	G,TLC	—	Cool, 4°C, 0.008% $Na_2S_2O_3$, store in dark	40 days after extraction
PCBs acrylonitrite	G,TLC	—	Cook, 4°C	40 days after extraction
Nitroaromatics and isophorone	G,TLC	—	Cool, 4°C, 0.008% $Na_2S_2O_3$, store in dark	40 days after extraction
Polynuclear aromatic hydrocarbons	G,TLC	—	Cool, 4°C, 0.008% $Na_2S_2O_3$, store in dark	40 days after extraction
Haloethers	G,TLC	—	Cool, 4°C, 0.008% $Na_2S_2O_3$	40 days after extraction
Chlorinated hydrocarbons	G,TLC	—	Cool, 4°C	40 days after extraction
TCDD	G,TLC	—	Cool, 4°C, 0.008% $Na_2S_2O_3$	40 days after extraction
Pesticide Tests				
Pesticides	G,TLC	—	Cool, 4°C, pH =5 to 9	40 days after extraction
Radiological Tests				
Alpha, beta and radium	P,G	—	HNO_3 to pH<2	6 months

Abbreviations:
P = Polyethylene
G = Glass
BT = Bottle and top
TLS = Teflon-lined septum
TLC = Teflon-lined cap

The Top Ten

things to do to avoid problems with an onsite system.

1. Do not dispose of fats, greases or cooking oils down the household drains.

2. Do not use a garbage disposal, put coffee grounds, meat bones or other food products that are difficult to biodegrade, down the drain.

3. Do not dispose of household cleaning fluids down the drain and use disinfectants sparingly.

4. Do not dispose of automotive fluids, such as gas, oil, transmission or brake fluid, greases or antifreeze down any drains.

5. Do not dispose of or rinse any containers from pesticides, herbicides or other potentially-toxic substances down any drain.

6. Do not dispose of any non-biodegradable substances or objects, such as cigarette butts, disposable diapers, feminine products (particularly, tampons).

7. Minimize water usage. Do not run water continuously while rinsing dishes or thawing frozen food products. Consider limiting toilet flushes.

8. Run only full loads when using a washing machine or dishwasher. Try to stagger use of the washing machine (i.e. Do not run six loads on Monday and none the other days.).

9. Do not use chemicals to "start up" or "clean" your system. They are unnecessary and may actually harm the system or the groundwater.

10. Do not connect any "clear water" sources, such as footing and foundation sump pumps to the sewage system.

Courtesy of Roman Kaminsky, Wisconsin Department of Commerce, Stevens Point, WI

Glossary

A horizon: The horizon formed at or near the surface, but within the mineral soil, having properties that reflect the influence of accumulating organic matter or elevation, alone or in combination.

absorption: The process by which one substance is taken into and included within another substance, as the absorption of water by soil or nutrients by plants.

activated sludge process: A biological wastewater treatment process in which a mixture of wastewater and activated sludge is agitated and aerated. The activated sludge is subsequently separated from the treated wastewater (mixed liquor) by sedimentation and wasted or returned to the process as needed.

adsorption: The increased concentration of molecules or ions at a surface, including exchangeable cations and anions on soil particles.

aerobic: Using molecular oxygen. Growing or occurring only in the presence of molecular oxygen, such as aerobic organisms.

aggregate, soil: A group of soil particles cohering so as to behave mechanically as a unit.

anaerobic: The absence of molecular oxygen. Growing in the absence of molecular oxygen (such as anaerobic bacteria).

anaerobic process: An anaerobic waste treatment process in which the microorganisms responsible for waste stabilization are removed from the treated effluent stream by sedimentation or other means, then returned to the process to enhance the rate of treatment.

bacteria: Single-celled microorganisms that feed upon and consequently degrade organic matter. Some bacteria are able to cause diseases in humans and animals.

B horizon: The horizon immediately beneath the A horizon characterized by a higher colloid (clay or humus) content, or by a darker or brighter color than the soil immediately above or below, the color usually being associated with the colloidal materials. The colloids may be of alluvial origin, as clay or humus; they may have been formed in place or they may have been derived from a texturally layered parent material.

biochemical oxygen demand (BOD): Measure of the concentration of organic impurities in wastewater. The amount of oxygen required by bacteria while decomposing organic matter under aerobic conditions, expressed in mg/l.

biological mat: A layer of microorganisms and inert organic materials which forms on the infiltrative surface of soil and provides treatment of the septic tank effluent.

blackwater: Liquid and solid human body waste and the carriage waters generated through toilet usage.

bulk density, soil: The mass of dry soil per unit bulk volume. The bulk volume is determined before drying to constant weight at 105°.

C horizon: The horizon that normally lies beneath the B horizon but may lie beneath the A horizon, where the only significant change caused by soil development is an increase in organic matter, which produces an A horizon. In concept, the C horizon is unaltered or slightly altered parent material.

calcareous soil: Soil containing sufficient calcium carbonate (often with magnesium carbonate) to effervesce visibly when treated with cold O.1N hydrochloric acid.

capillary attraction: A liquid's movement over, or retention by, a solid surface, due to the interaction of adhesive and cohesive forces.

cation exchange: The interchange between a cation in solution and another cation on the surface of any surface-active material, such as clay or organic colloids.

cation-exchange capacity: The sum total of exchangeable cations that a soil can adsorb; sometimes called total-exchange, base-exchange capacity, or cation-adsorption capacity. Expressed in milliequivalents per 100 grams or per gram of soil (or of other exchanges, such as clay).

chemical oxygen demand (COD): A measure of the oxygen equivalent of that portion of organic matter that is susceptible to oxidation by a strong chemical oxidizing agent.

clarifiers: Settling tanks. The purpose of a clarifier is to remove settleable solids by gravity, or colloidal solids by coagulation following chemical flocculation; will also remove floating oil and scum through skimming.

clay: (1) A soil separate consisting of particles <0.002 mm in equivalent diameter. (2) A textural class.

clay mineral: Naturally occurring inorganic crystalline or amorphous materials found in soils and other earthy deposits, the particles being predominantly <0.002 mm in diameter. Largely of secondary origin.

coarse texture: The texture exhibited by sands, loamy sands, and sandy loams except very fine sandy loams.

coliform-group bacteria: A group of bacteria predominantly inhabiting the intestines of man or animal, but also occasionally found elsewhere. Used as an indicator of human fecal contamination.

columnar structure: A soil structural type with a vertical axis much longer than the horizontal axes and a distinctly rounded upper surface.

conductivity, hydraulic: As applied to soils, the ability of the soil to transmit water in liquid form through pores.

consistence: (1) The resistance of a material to deformation or rupture. (2) The degree of cohesion or adhesion of the soil mass.

Terms used for describing consistence at various soil moisture contents are:

> **wet soil**: Nonsticky, slightly sticky, sticky, very sticky, non-plastic, slightly plastic, plastic, and very plastic.
>
> **moist soil**: Loose, very friable, friable, firm, very firm, and extremely firm.
>
> **dry soil**: Loose, soft, slightly hard, hard, very hard, and extremely hard.
>
> **Cementation**: Weakly cemented, strongly cemented, and indurated.

crumb: A soft, porous, more or less rounded ped from 1 to 5 mm in diameter.

crust: A surface layer on soils, ranging in thickness from a few mm to perhaps as much as an inch, that is much more compact, hard, and brittle when dry, than the material immediately beneath it.

denitrification: The biochemical reduction of nitrate or nitrate to gaseous molecular nitrogen or an oxide of nitrogen.

digestion: The biological decomposition of organic matter in sludge, resulting in partial gasifiction, liquefaction, and mineralization.

disinfection: Killing pathogenic microbes on or in a material without necessarily sterilizing it.

disperse: To break up compound particles, such as aggregates, into the individual component particles.

dissolved oxygen (DO): The oxygen dissolved in water, wastewater, or other liquid, usually expressed in milligrams per liter (mg/l), parts per million (ppm), or percent of saturation.

dissolved solids: Theoretically, the anhydrous residues of the dissolved constituents in water. Actually, the term is defined by the method used in determination.

effluent: Sewage, water, or other liquid, partially or completely treated or in its natural state, flowing out of a reservoir, basin, or treatment plant.

effective size: The size of grain such that 10% of the particles by weight are smaller and 90% greater.

eutrophication: Accelerated aquatic plant growth in waterbody due to high concentrations of nutrients. Deficiency of dissolved oxygen in the waterbody often results from eutrophication.

evapotranspiration: The combined loss of water from a given area, and during a specified period of time, by evaporation from the soil surface and by transpiration from plants.

extended aeration: A modification of the activated sludge process which provides for aerobic sludge digestion within the aeration system.

filtrate: The liquid which has passed through a filter.

fine texture: The texture exhibited by soils having clay as a part of their textural class name.

food to microorganism ratio (F/M): Amount of BOD applied to the activated sludge system per day per amount of MLSS in the aeration basin, expressed as lb BOD/d/lb MLSS.

graywater: Wastewater generated by water-using fixtures and appliances, excluding the toilet and possibly the garbage disposal.

groundwater: Water found in pore spaces in the subsurface below the water table.

groundwater mound: The elevated groundwater level produced by introducing septic tank effluent into the soil.

hardpan: A hardened soil layer, in the lower A or in the B horizon, caused by cementation of soil particles with organic matter or with materials such as silica, sesquioxides, or calcium carbonate. The hardness does not change appreciably with changes in moisture content, and pieces of the hard layer do not slake in water.

hydraulic conductivity: The ability of the soil to transmit liquids through pore spaces in a specified direction, e.g., horizontally or vertically.

impervious: Resistant to penetration by fluids or by roots.

infiltrative surface: The area of the soil that is exposed to effluent.

influent: Water, wastewater, or other liquid flowing into a reservoir, basin, or treatment plant.

intermittent filter: A natural or artificial bed of sand or other fine-grained material to the surface of which wastewater is applied intermittently in flooding doses and through which it passes; opportunity is given for filtration and the maintenance of an aerobic condition.

ion: A charged atom, molecule, or radical, the migration of which affects the transport of electricity through an electrolyte or, to a certain extent, through a gas. An atom or molecule that has lost or gained one or more electrons; by such ionization it becomes electrically charged. An example is the alpha particle.

ion exchange: A chemical process involving reversible interchange of ions between a liquid and a solid but no radical change in structure of the solid.

leaching: The removal of materials in solution from the soil.

loading rate: The amount of effluent that is applied to the soil, usually in liters/square meter or gallons/square feet.

Long Term Acceptance Rate: The stable rate of effluent acceptance through a biological mat.

mapping unit: A soil or combination of soils delineated on a map and, where possible, named to show the taxonomic unit or units included. Principally, mapping units on maps of soils depict soil types, phases, associations, or complexes.

medium texture: The texture exhibited by very fine sandy loams, loams, silt loams, and silts.

mineral soil: A soil consisting predominantly of, and having its properties determined by, mineral matter. Usually contains <20 percent organic matter, but may contain an organic surface layer up to 30 cm thick.

mineralization: The conversion of an element from an organic form to an inorganic state as a result of microbial decomposition.

mineralogy, soil: In practical use, the kinds and proportions of minerals present in soil.

mixed liquor suspended solids (MLSS): Suspended solids in a mixture of activated sludge and organic matter undergoing activated sludge treatment in the aeration tank.

mottling: Spots or blotches of different color or shades of color interspersed with the dominant color.

nitrate (NO_3^{-1}): the most highly oxidized form of inorganic nitrogen and a frequent groundwater contaminant generated by septic systems.

nitrification: The biochemical oxidation of ammonium to nitrate.

onsite subsurface sewage system: A passive or active method for treating and disposing of wastewater into the soil/groundwater system.

organic nitrogen: Nitrogen combined in organic molecules such as proteins, amino acids.

organic soil: A soil which contains a high percentage (>15 percent or 20 percent) of organic matter throughout the soil.

particle size: The effective diameter of a particle usually measured by sedimentation or sieving.

particle-size distribution: The amounts of the various soil separates in a soil sample, usually expressed as weight percentage.

pathogenic: Causing disease. "Pathogenic" is also used to designate microbes which commonly cause infectious diseases, as opposed to those which do not uncommonly or never.

ped: A unit of soil structure such as an aggregate, crumb, prism, block or granule, formed by natural processes (in contrast with a clod, which is formed artificially).

percolation: The flow or trickling of a liquid downward through a contact or filtering medium. The liquid may or may not fill the pores of the medium.

permeability, soil: The ease with which gases, liquids, or plant roots penetrate or pass through soil.

pH: A term used to describe the hydrogen-ion activity of any system.

phosphorus: An element associated with the accelerated eutrophication of fresh water bodies. In many case, phosphorus is the limiting nutrient in water bodies.

platy structure: Soil aggregates that are developed predominantly along the horizontal axes; laminated; flaky.

recirculating sand filter: A biological and physical treatment process consisting of a bed of sand to which effluent from the septic tank is discharged and recirculated prior to disposal

redoximorphic features: specific coloration of the soil that indicates alternating periods of water saturation, thus reducing conditions.

saturated zone: The area below the water table where the soil pores are fully saturated with water.

settleable solids: That matter in wastewater which will not stay in suspension during a preselected settling period, such as one hour, but either settles to the bottom or floats to the top.

silt: (1) A soil separate consisting of particles between 0.05 and 0.002 mm in diameter. (2) A soil textural class.

single-grained: A nonstructural state normally observed in soils containing a preponderance of large particles, such a sand. Because of a lack of cohesion, the sand grains tend not to assemble in aggregate form.

site elevation: The process for determining if a site is capable of providing treatment and disposal of wastewater.

slope: Deviation of a plane surface from the horizontal.

soil horizon: A layer of soil or soil material approximately parallel to the land surface and differing from adjacent genetically related layers in physical, chemical, and biological properties such as color, structure, texture, consistence, pH, etc.

soil map: A map showing the distribution of soil types or other soil mapping units in relation to the prominent physical and cultural features of the earth's surface.

soil morphology: The physical form of the soil, evidenced by a soil profile showing the kinds, thickness, and arrangement of the horizons in the profile, and by the texture, structure, consistence, and porosity of each horizon.

soil separates: Groups of mineral particles separated on the basis of a range in size. The principal separates are sand, silt, and clay.

soil series: The basic unit of soil classification, and consisting of soils which are essentially alike in all major profile characteristics, although the texture of the A horizon may vary somewhat. See soil type.

soil solution: The aqueous liquid phase of the soil and its solutes consisting of ions dissociated from the surfaces of the soil particles and of other soluble materials.

soil structure: The combination or arrangement of individual soil particles into definable aggregates, or peds, which are characterized and classified on the basis of size, shape, and degree of distinctness.

soil survey: The systematic examination, description, classification, and mapping of soils in an area.

soil texture: The relative proportions of the various soil separates in a soil.

soil type: for mapping soils, a subdivision of a soil series based on differences in the texture of the A horizon.

soil water: A general term emphasizing the physical rather than the chemical properties and behavior of the soil solution.

solids: Material in the solid state.

total suspended (TSS): The solids remaining as residue after water has been evaporated from a sample (freshwater or wastewater).

dissolved (TDS): Solids present in solution.

suspended: Solids physically suspended in water, sewage, or other liquids. The quantity of material deposited when a quantity of water, sewage, or liquid is filtered through an asbestos mat in a Gooch crucible.

volatile: The quantity of solids in water, sewage, or other liquid lost on ignition of total solids.

subsoil: In general concept, that part of the soil below the depth of plowing.

tension, soil water: The expression, in positive terms, of the negative hydraulic pressure of soil water.

textural class, soil: Soils grouped on the basis of a specified range in texture. In the United States, 12 textural classes are recognized.

tight soil: A compact, relatively impervious and tenacious soil (or subsoil), which may or may not be plastic.

topography: The general shape of the ground's surface at a particular site.

topsoil: The layer of soil in cultivation; The A horizon. Fertile soil material used to topdress roadbanks, gardens, and lawns.

unsaturated flow: The movement of water in a soil which is not filled to capacity with water.

unsaturated zone: The area above the water table where the soil pores are not fully saturated, although some water may be present (vadose zone).

water table: That level in saturated soil where the hydraulic pressure is zero.

water table, perched: The water table of a discontinuous saturated zone in a soil.

INDEX